Leveling the Playing Field

Leveling the Playing Field

The Democratization of Technology

Rod Scher

Guilford, Connecticut

An imprint of Rowman & Littlefield

Distributed by NATIONAL BOOK NETWORK

British Library Cataloguing in Publication Information Available

Library of Congress Cataloging-in-Publication Data Available

ISBN 978-1-4422-3926-5 (hardback)
ISBN 978-1-4422-3927-2 (e-book)

™ The paper used in this publication meets the minimum requirements of American National Standard for Information Sciences—Permanence of Paper for Printed Library Materials, ANSI/NISO Z39.48-1992.

This one is for
William "Papa Bill" Sachs
Always the best father he could be, even when he didn't have to be—
and always the friend and guide we needed, even when
we didn't realize that we needed one.

Technology, when misused, poisons air, soil, water, and lives. But a world without technology would be prey to something worse: the impersonal ruthlessness of the natural order, in which the health of a species depends on relentless sacrifice of the weak.

—*New York Times*, August 29, 1986

Table of Contents

Acknowledgments

This book owes a great deal to the advice, input, and suggestions of many people, many of them experts in the various fields touched on in the book, and also to those who were ready to lend a hand when it came time to edit, review, fact-check, and help the author organize what turned out to be a mammoth undertaking. Almost every source I contacted was extremely supportive and interested in helping the project move forward, and I thank them all. Some, though, went above and beyond, answering dozens of (occasionally stupid) follow-on questions, sitting through multiple interviews, and putting me in touch with colleagues who could also contribute their expertise. Those folks deserve an extra tip of the hat, along with my deep gratitude.

So, in no particular order (other than alphabetical), I'd like to offer my most sincere thanks to the following people: Dr. Edoé Djimitri Agbodjan, Dr. Steve Buhler, Dr. Celeste Campos-Castillo, Chad Denton, Chris Hadnagy, Dr. Sean Hess, Dr. Ryan Kirkbride, Alex Lifschitz, Jim O'Gorman, Zoe Quinn, Dr. Antonio Sagona, Dr. Daniel Schneider, Dr. Terry Twomey, and Dulcie Wilcox.

A special note of thanks goes to my research assistant, Jaime Detour, who read, fact-checked, and dug up obscure bits of information—and who somehow managed to organize the permissioning of many dozens of photos and images.

And finally, a huge thank-you to my wife, Lesley. I'm not sure what I did to deserve you, but whatever it was, I'm so glad I got it right that one time.

Introduction

1983: Looking Back

In general, it wasn't the best of years: 1983 began with the largest part of Russia's Cosmos 1402 satellite plummeting into the Indian Ocean. The four-ton machine contained its very own nuclear reactor and about a hundred pounds of enriched uranium. The spacecraft was so radioactive, and its trajectory so wild and unplanned, that the entire world sighed in relief when the battered and charred remnants of the satellite, including the intensely radioactive U-235 core, splashed down in the South Atlantic a few weeks later.

But the year had only just begun, and more trouble with the USSR loomed. In March, President Reagan, in a speech to a group of influential clergymen, called the USSR an "evil empire." In June, the Soviets, in protest of the United States' recent deployment of Pershing missiles in West Germany, simply walked out of disarmament talks taking place in Geneva. Then, in late August, when an unarmed South Korean airliner strayed into Soviet airspace, it was shot down by an Su-15 fighter. All 269 passengers and crew aboard were killed, including US Congressman Lawrence McDonald.

It was certainly a dodgy year politically, and that wasn't the extent of it: Mother Nature was also in high dudgeon.

During the same year, a March earthquake in Colombia killed some 5,000 people and caused $380 million in damage, a California flood killed 13, and a tornado and hurricane in Texas combined to kill 29 people and cause billions of dollars' worth of destruction. (Meanwhile, Yankee outfielder Dave Winfield struck back at Mother Nature, accidentally killing a seagull while warming up before the fifth inning of an otherwise unremarkable baseball game in Toronto.)

Of course, Mother Nature wasn't the only one intent on destruction. As always, humans continued to conspire to ruin their own planet.

In February, the EPA decided that the town of Times Beach, Missouri, was so polluted (the result of flooding and a misadvised contractor who had spread dioxin-tainted oil on the town's unpaved roads to keep the dust down) that the government agency had no choice but to *buy* the town and then pay to have every one of its more than two thousand inhabitants relocated. The town was demolished in 1992, and today remains an overgrown but otherwise featureless plain bisected by Interstate Highway 44. Automobiles zip past at 70 miles per hour; few of the drivers and passengers realize that they're passing through what was once a small town.

Then, in October, the United States invaded Grenada, an island nation with a population of about 91,000—that is, slightly more people than currently reside in Trenton, New Jersey. An unsurprisingly swift and unequivocal victory resulted, and the victors celebrated by awarding more than 5,000 medals to the 7,000-plus soldiers, sailors, and airmen who had participated in the conflict.

More calamity and disaster loomed, of course, including the introduction of both the minivan and the Chicken McNugget, but after a while, the sordid recap begins to wear, so perhaps it's time to move on.

The Year in Technology

On the other hand, as bad as 1983 was in other respects, it turned out to be an excellent year—perhaps even a seminal one—for technology. The year saw the introduction of Microsoft's original version of Word (possibly called Multi-Tool Word at the time, though opinions about that differ), as well as the debut of Apple's IIe and Lisa computers—the first of which had helped begin one revolution and the second of which surely portended another, by virtue of its now-ubiquitous graphical user interface.

In that year we also welcomed the CD-ROM, the first mobile cellular phone (a $4,000, three-pound behemoth), and Mitch Kapor's Lotus 1-2-3—a spreadsheet program that would steal the "killer app" crown from Dan Bricklin's VisiCalc almost overnight.

Also in 1983, Compaq gave us the first real PC clone and Sony introduced consumer-level videotape—in what would turn out to be the short-lived Beta format, of course.

Finally, it was a year of two more technology firsts, though these may have been on opposite ends of the spectrum: It was the year in which the Advanced Research Projects Agency Network (ARPANET) converted its networking topology, arguably marking the moment that the modern Internet was born. It was also the year in which Fred Cohen, a USC computer science student working on his dissertation, designed and defined a computer program that could affect other programs by modifying them to include a copy or near-copy of itself; Fred had just invented—or at least formalized—the computer virus. (He would go on to pioneer many antivirus defense techniques and technologies, and write dozens of papers and books on the subject.)

3D PRINTING: A FACTORY IN A BOX

But one of the most important and ultimately among the most far-reaching technological inventions of 1983 was one that largely escaped notice at the time. It was in 1983 that a middle-aged engineer and parts designer named Chuck Hull began looking for a way to quickly and inexpensively prototype parts. At the time, this was, as Chuck has noted someone understatedly, "a tedious process." What he meant was that designing and fabricating a prototype part took weeks or months, and could cost a small fortune—all to create a part that might in fact turn out to be useless.

Chuck realized that one could instead build a part in layers—a process that has come to be called *additive manufacturing*—by carefully spraying a thin layer of plastic or metal in the desired shape, and then spraying more layers on top of that first one, building up the item one layer at a time until one had a fully formed part.

It was a brilliant intellectual leap, but much work remained: A "printer" such as Chuck envisioned did not yet exist; he had to invent one. The software that translated the part's specifications and fed them to the printer also did not exist; he had to invent that, too. He even had to invent the STL file format that the software read in order to begin translating the item specs into instructions it could send to the printer.

In the end, 3D printing (this form of which is technically called *stereolithography*), though invented in 1983, would take several years to mature. (Hull's patent, #4575330, was filed on August 8, 1984. He now holds more than sixty patents related to the process.) In 1986, Hull founded 3D Systems in Valencia, California. The company now has hundreds of employees and dozens of offices around the country.

Large-scale 3D printing began as, and to a certain extent, remains, the domain of either companies that can afford expensive commercial printers to create parts made of metal or plastic—think Boeing, which used the process to create much of its recent drone fleet; and NASA, which is experimenting with the 3D printing of replacement parts in space—or of talented hobbyists who enjoy tinkering with 3D printers, or even building them from scratch. However, this is changing.

The technology has recently begun trickling down. One can now buy, for between $500 and $3,000, 3D printers capable of printing multicolored parts in multiple types of plastic. These are quite suitable for making small parts for experimentation, or for making such things as hooks, stands, bowls, and other such widgets. (Using one of the larger printers, one can even print the body of a fairly decent-sounding electric guitar.)

It's not overstating things to say that this technology could easily turn out to be the future of manufacturing, and may in fact herald a drastic and relatively sudden transformation in global economics. What do you suppose will happen to manufacturing—and to its allied vendors and suppliers, including companies that sell warehouse equipment or shipping services—when the manufacture of an appreciable percentage of goods can be done at home, or in a garage or small office? The impact could be staggering.

And one man is behind a movement dedicated to making that very thing happen. Adrian Bowyer, a British engineer and mathematician who formerly taught at Bath University, sees the spread of 3D printing as a form of empowerment, and he is determined to see that its availability helps to level the playing field.

Bowyer is probably not as widely known as we would expect, perhaps because he publishes papers with such titles as "The Influence of Degree-of-Branching and Molecular Mass on the Interaction between Dextran and Concanavalin A in Hydrogel Preparations Intended for Insulin Release," which ended up in a 2012 issue of the *European Journal of Pharmaceutics and Biopharmaceutics*, and was consequently read by almost no one. In fact, among the many publications he lists in his CV, only three have anything to do with 3D printing. He is clearly a man of many talents and interests.

But one of his interests—by most accounts, an all-consuming one—is RepRap.

RepRap is the world's first low-cost, open-source 3D printer. People are free (in fact, encouraged) to copy it, to improve it, and to ensure that it spreads across the globe. Plans and specs are readily available. In fact, Bowyer *designed* it to be copied. RepRap is not simply a 3D printer; it's a printer that can print copies of itself. (Which perhaps recalls the virus definition noted above, but in a somewhat more positive way.) In fact, Bowyer's egalitarian intent is to create a printer that allows anyone who builds or buys one to create more printers for his friends, neighbors, and colleagues. Bowyer envisions a world in which anyone who wants a 3D printer can have one.

Democratizing Technology

This book is about tools such as these, technologies that "trickle down" to the rest of us, those that were once the domain of the wealthy and powerful—and which therefore tended to make them even *more* wealthy and powerful. Now, though, these technologies—from books to computers to 3D printing and beyond—have become part of a common toolkit, one accessible to almost anyone, or at least to many more than had heretofore had access. This is what happens with most technologies: They begin in the hands of the few, and they end up in the hands of the many. Along the way, they sometimes transform the world.

If Bowyer has his way, RepRap could turn out to be the ultimate example of a democratized, transformative technology. Will it be disruptive? Certainly. If RepRap succeeds, people will be thrown out of work, and the entire focus of manufacturing could shift such that in many cases an item's point of sale is also the place at which it was manufactured. (This idea is not at all new, of course; it's how many artisans' guilds and other workshops operated for centuries: Products from pottery to horseshoes to fishing rods were often sold at—or right outside of—the place of manufacture.)

The disruption will, as all disruptions do, cause problems for entrenched suppliers and manufacturers, but there is simply no avoiding it. As an analogous example, consider the rise of e-commerce. Bemoaning the Internet-fueled fate of so-called brick-and-mortar shops is simply useless; that genie is completely out of the bottle, and cannot be put back. Rightly or wrongly, for better or worse, the Internet and all of its benefits and ills are here to stay.

Retailers, the sharp ones, are learning how to coexist with and even profit from the Internet, rather than wasting their time fighting it, complaining about it, or attempting to somehow rein it in.

The smart money is on companies that see Internet sales as an opportunity, and which thus look for ways to take advantage of that opportunity instead of lamenting the fact that the world has changed. The spread of 3D printing—especially if RepRap or some similar movement catches on—will result in a similar seismic shift. We'll still need companies to make (and warehouse and ship) products, of course, but many items will be printed in homes and small shops. In such cases, the real "product" then becomes the set of specifications contained in the STL or other such file: Instead of selling a bowl that you crafted or machines that you built, a portion of the new workforce will be selling, in effect, the *plans* for such a bowl, the information about how to make that machine, and also the new software that can read such files and then transmit instructions to an ever-evolving breed of 3D printers. Essentially, the twenty-first-century entrepreneur sells *information.*

Is there a downside? Absolutely, and it's more than just workforce disruption. For instance, if 3D printing becomes more accessible, people will use it—are already using it—to print dangerous objects, such as guns. But then every manufacturing technology can be used to make potentially dangerous objects, from knives to guns to bricks. At the moment, lathes are much more dangerous than 3D printers, but no one has yet outlawed the former.

Technologies, after all, reflect human nature, and it is our nature to build, to create. Humans make things; it's part of what makes us human. And, as Adrian Bowyer has noted, though we can certainly build dangerous or destructive items, "People make more benign things than they make harmful things."

In the end, whether or not we favor the 3D printing revolution is largely irrelevant. Like the Internet (and like books and writing itself), the technology is here to stay—at least until such time as it's replaced by still more advanced technology. It's a powerful tool that industry has used for years to create products and prototypes; now, in the hands of the masses, it empowers those masses. The result will not necessarily be a perfect world; it may not even be a better world. But it will be a changed one.

We'll have an opportunity to discuss 3D printing in more detail in chapter 8, but as we begin our look at how technology becomes democratized—and how it then itself becomes a democratizing force—it may be useful to first go back to the beginning. The *very* beginning. And perhaps the first and most transformative of these technological shifts came about with the discovery of—or more properly, with the harnessing of—that most elemental of forces: fire.

Chapter One

Foundations: Fire and Stone

It is only when science asks why, instead of simply describing how, that it becomes more than technology. When it asks why, it discovers Relativity. When it only shows how, it invents the atom bomb, and then puts its hands over its eyes and says, "My God what have I done?"

—Ursula K. Le Guin

The Giant in the Cave

Once there was a giant who lived in a cave. He was by all accounts a fairly gentle giant, and a peaceful, hardworking one. He mined the cave for guano—bat and owl droppings, mainly—which he sold as fertilizer to the farmers in the area around the cave.

After several years, the giant married. Soon, his wife gave birth to a daughter whom they called Girlie, a name that sounds unimaginative to us, but which was common enough in that time and at that place.

The giant, whose name was Pieter, lived in the cave with his family for a number of years. They cooked and slept not too far from the mouth of the cave, where a smooth stone floor had been laid to help level the area and to support their stove and their belongings. The stone also covered and protected the family from the dirt that was the natural floor of the cave. When the weather was harsh, Pieter and his family sometimes sheltered their livestock farther back in the cave, for the cave was quite large, with a depth of several hundred feet; there were parts of the cave that Pieter's family rarely visited, except to dig for the soft, loamy, rich soil that provided their livelihood.

But, while it sounds like one, this was no fairy tale. Pieter Bosman was a real person, though a very large one—a resident of the Kuruman Hills in Northern Cape, South Africa. He had inherited the cave from his father, P. E. Bosman, in the early 1900s. The elder Bosman had lived in the cave with *his* wife and children while building a farmhouse. Many of the dozen or so Bosman children had been born in that very cave. When the house was complete, Pieter elected to stay in the cave, while his father, mother, and the remaining Bosman children moved into the house. Perhaps Pieter, the eldest son, sought some peace and quiet, both of which would have been in short supply in a small house filled to the brim with people, many of them small, noisy, and perhaps a bit rambunctious.

But without realizing it, Pieter was digging up and selling history. Much of the "guano" he mined was actually made up of millions of years' worth of sediment in which resided the history—and the prehistory—of a people. Inscribed in the dirt and rock of the cave, sometimes literally, was the story of man. The giant—he wasn't a true, fairy-tale giant, but he weighed some four hundred pounds and stood six foot, five inches tall, at a time when the average man was well under six feet in height—kept uncovering shards of pottery, bone chips, and what looked suspiciously like Stone Age tools. Eventually, scientists came to see what was in the cave, and soon the place was crawling with archaeologists, anthropologists, and other -ologists, all of whom were astounded at the richness of the find: Inside the protective cave, preserved in layers of sediment that could be fairly accurately dated, was two million years of history. The layers told a story that stretched from the cave's relatively recent tribal inhabitants all the way back to before true humans even existed.

And at about the one-million-year layer inside what was now called the Wonderwerk Cave—and not discovered and confirmed until the mid-2000s—was something truly incredible. Embedded in the soil and stone of Wonderwerk was ash that scientists determined had come not from a forest fire or from burning tundra that had washed or blown into the cave from outside, but from fires maintained within the cave; these were fires *managed by man*. Fires used to heat, to cook, to extend the day, and to keep away wild beasts. It was proof that man's forebears, *Homo erectus*, most likely, had controlled fire for a million years, at least two hundred thousand years longer than had been supposed.

This was the earliest confirmed use of controlled fire, and all this time it had been in P. E. Bosman's cave, a place where the family lived and slept and, perhaps somewhat ironically, cooked their meals. (We tend to think of *Homo erectus*, when we think of him at all, in the same vein as *Neanderthal*—as a brutish, unrefined subhuman, more animal than man. But keep in mind that it was—as near as we can tell—*Homo erectus* who learned to control fire. And also that, as a species, *Homo erectus* survived for more than five times as long as has our own species so far. He was supremely adapted for his environment, and it was he who migrated out of Africa and began populating Earth. Without *erectus*, the face of the planet would be very different, and we would most likely not be here to remark on that difference.)

CREATURES OF FIRE

In many ways, fire can be said to have made us human, and the transition began quite early.

"The significance of fire use," says University of Melbourne anthropologist Terry Twomey, "is that it probably predates complex technologies like bows, baskets, pottery, spear throwers, nets, and hafted tools by hundreds of thousands of years."

In fact, agriculture and metallurgy, two of our biggest technological developments (developments that led in turn to other important tools and technologies) were fire-dependent. After all, if you can't cook bread, there's really no agriculture. "Every cooking method is also a direct result of using fire," says Dr. Twomey. "Once we had mastered fire, everything was possible, and most of the tools we have are a result of this."

Harvard University anthropologist Richard Wrangham, meanwhile, has caused something of a stir by suggesting that fire is quite *literally* what made us human. His "cooking theory" posits that cooked food allowed prehistoric man to extract more calories from his food than before, calories that his growing brain needed to develop into the cognitive powerhouse of an organ that it has become. It also, says Wrangham, allowed us to evolve the smaller teeth and guts that made us more suited to living upright on the tundra than in trees.

Wrangham believes, in fact, that we'll eventually discover that man was controlling (though not creating) fire even earlier—perhaps as long as two million years ago.

Dr. Terry Twomey has studied the effects of fire on *Homo erectus*. This model of a female *Homo erectus* is on display at the Smithsonian Museum of Natural History in Washington, DC Dr. Twomey believes that the control of fire required (or helped initiate) even greater cognitive growth than did the creation of other complex tools. *Image licensed under the Creative Commons Attribution-Share Alike 2.0 Generic license.*

Asked about comparing his work to Wrangham's, Terry Twomey says, "He focuses mainly on biology to argue his case, whereas I focus on behavior. I focus more on the social and cognitive implications of cooking and other fire-related behaviors. I rely a lot on Wrangham's and Ofek's claims as premises in my argument, but I extend on their claims in terms of the cognitive implications of cooking and using fire without being able to make it." (Binghamton University anthropologist Dr. Haim Ofek agrees with Wrangham, holding that "the relatively large brains in humans . . . could not have been achieved without a shift to a high-quality diet.")

Of course, just about any technology both drives and is driven by cognitive evolution, and it's not always clear which is cause and which is effect.

Says Twomey: "Fire use, in conjunction with more-complex foraging behaviors and tool-making, would have driven our cognitive evolution. However, I think fire use is particularly significant because it makes greater and more significant demands on cognition than early human tool use and foraging. The exact timing for the emergence of fire use is important in this regard, as I am assuming a long prehistory. While this is still debated, the evidence that regular fire use was happening at least one and a half million years ago is getting stronger all the time. This places fire use at a critical stage in the evolution of human society and cognition, long before the more-complex technologies I mentioned appear."

Social Considerations

As Twomey says, the control of fire did more than merely affect our physical evolution. It also may have helped to determine who we were socially, and in doing so, influenced the course of man's social development through the ages. For example, if nothing else, control of fire demonstrated an ability to plan ahead, because it required gathering and storing fuel; it also demanded a fairly sophisticated level of cooperation that would have been necessary in order to provision a fire and keep it burning.

Even language may have evolved partly because of fire.

"Planning ahead to secure future mutual benefits was a critically important development," says Dr. Twomey. "However, this is probably not the most impressive aspect of MP [Middle Pleistocene, an age that began around 781 to 126 thousand years ago] human life. I think the advanced communication skills that fire use implies were perhaps the most significant, although these are closely related to our capacity to cooperate and plan ahead. MP humans probably had a pretty advanced language capable of making plans with others and passing on social norms. The influence of fire on the evolution of language is an interesting and important problem in my research. Fire use could have facilitated and stabilized key communication skills in human societies that were necessary for language to evolve."

When one thinks about the effects of fire—that is, those beyond "merely" providing cooked food—one is struck by the immense breadth and impact of those effects. This is truly a foundational technology, a tool without which few subsequent tools could have been invented, and one whose impact is so far-reaching that thoroughly evaluating and appreciating that impact is almost impossible.

AROUND THE CAMPFIRE

First, consider that a fireplace, hearth, or campfire is and always has been a place for socialization, for discussion, even for argument. It's where people gather now, and of course it was where people gathered those hundreds of thousands of years ago. It may have been the very place they gathered to discuss the pooling of resources that was made necessary by the fire itself—Who keeps it going? Who gathers wood? When? Who helps? What do we do about those who refuse to help?—or by the hunting that was given new emphasis by this amazing technology that enabled our forebears to cook the food they hunted.

Learning to use, maintain, and control fire—and then, somewhat later, to create it—assumes the existence of both technical tools and an ever-expanding repertoire of social behaviors aimed at providing a common good: Fire benefits everyone in the group, so those who maintained and carried fire were, we assume, valued. Pliocene and the later Pleistocene "fire marshals" would have been honored and perhaps cared for, some of their work taken on by other group members who desired to benefit from the fire that the fire maintainer would build (and share) that evening.

In fact, some anthropologists wonder whether this whole evolution of "home and hearth" might have led to interactions of a sort that might not have come about otherwise—or that might have been delayed for thousands of years. Perhaps storytelling evolved here. If so, what did that lead to? An entire culture made richer by the invention of (and the information communicated by) an oral tradition? The invention and sharing of myth? Is this where our origin stories began, around a flickering fire many, many thousands of years ago? If God needed to be invented—or His existence discovered or debated—what role might fire have thus played in that invention or discovery?

We still see today echoes of the socializing effects of fire.

Idaho-based archaeologist Sean Hess has studied timber-growing tribes in the Pacific Northwest, including their use of fire in pithouses. This is the interior of a pithouse used by the Sinixt nation in British Columbia. *Image licensed under the Creative Commons Attribution-Share Alike 2.0 Generic license.*

Archaeologist Sean Hess, based in Idaho, has studied the tribes of the Pacific Northwest. He notes that a communal fire—and communal buildings to house that fire—have long played a part in the tribes' cultural and social heritage.

"In the Pacific Northwest, one of the common prehistoric house forms was something called a 'pit house,'" says Hess. "These were usually circular in pattern, at least in the interior Pacific Northwest. They were often dug down two to four feet, and then there was a wooden framework that was built up inside of the circular pit house that allowed people to put up . . . vertical supports, and then they leaned a series of rafters onto those vertical supports and covered up the whole arrangement with mats, and then sometimes with a layering of earth. The centers of these pit houses were open, and in many cases that was the means of ingress into the structure. That's also where the fire was located, and so you have basically the whole living arrangement of people centered on where that fire was, and the fire was right in the middle of their communal lodge."

In effect, the tribes' winter homes were built, literally and figuratively, around a campfire.

It may be that fire did more than simply impact our biological evolution; it may have played an important role in the transmission of knowledge, of information, of culture. Fire may have created—or played an important part in creating—our society.

Safety and Productivity

The prehistoric African plain was not a safe place. Wild beasts roamed the land, and wolves had not yet evolved into the protective and helpful canine friends they are today. (The best guess is that wolves evolved into true dogs some twenty to thirty thousand years ago. Even there, though, fire may have played a part: Most researchers think that today's domesticated canines evolved from the first few beasts that were brave enough to approach the campfire for warmth and for food scraps.)

For the time being, man was alone, and a fire helped to keep the beasts away. It gave off a warmth that was pleasant, in some cases, perhaps lifesaving. But perhaps even more importantly, the fire offered protection. Think what it must have felt like to rest, to sleep soundly—perhaps for the first time—knowing that wild animals would be unlikely to venture into a cave or campground while a fire flickered nearby. Those of us who sleep cocooned in our soft sheets and blankets, warm and safe in our in heated homes, have little understanding of what it must have felt like to be at the mercy of wicked weather and wild beasts.

Fire did more than simply protect us and allow us to cook and preserve food—although these factors alone would have been beneficial in both the short and long terms. It also did something very important that we now take for granted: It extended the day. Instead of everyone going to sleep when the sun went down, now people could work, play, and socialize into the night.

The University of Utah's Polly Wiessner has studied the African Bushmen, latter-day counterparts of *Homo erectus* and other prehistoric hunter-gatherers. From her study of these tribesmen, she has noted that even

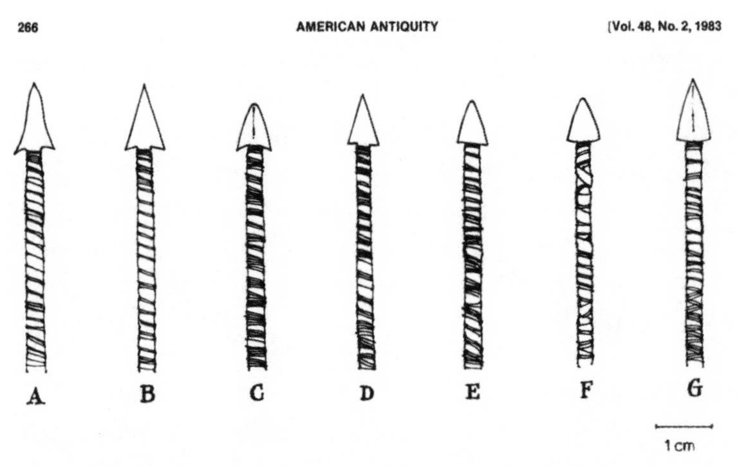

Dr. Polly Wiessner, of the University of Utah, has investigated the fireside conversations of the Ju/'hoansi Bushmen, comparing their behavior to prehistoric hunter-gatherers such as *Homo erectus*. In an earlier work (1983), she compared stylistic variability in projectile points among linguistically related southern African hunter-gatherers. *This image is licensed under Creative Commons Attribution-NoDerivs 3.0 Unported (CC BY-ND 3.0). See http://creativecommons.org/licenses/by-nd/3.0/.*

though each Bushman family may have had a fire, at night people often met at a communal hearth. Dr. Wiessner wonders what goes on today in what she calls "these fire-lit spaces," and, by extension, what went on all those thousands of years ago.

"Fire altered our circadian rhythms, the light allowed us to stay awake, and the question is, what happened in the fire-lit space? What did it do for human development?" she asks. In her study, "Embers of Society: Firelight Talk among the Ju/'hoansi Bushmen," Wiessner found that the evening conversations around the fire sometimes centered on many of the same topics that we (and presumably *Homo erectus*) might also have discussed: murder, extramarital affairs, disputes, and interactions with other groups. (The Bushmen also spent a fair amount of time discussing bush fires, being chased by animals, and fights over meat, topics with which one hopes most of us would have had little experience, but which might have been of interest to our ancestors.)

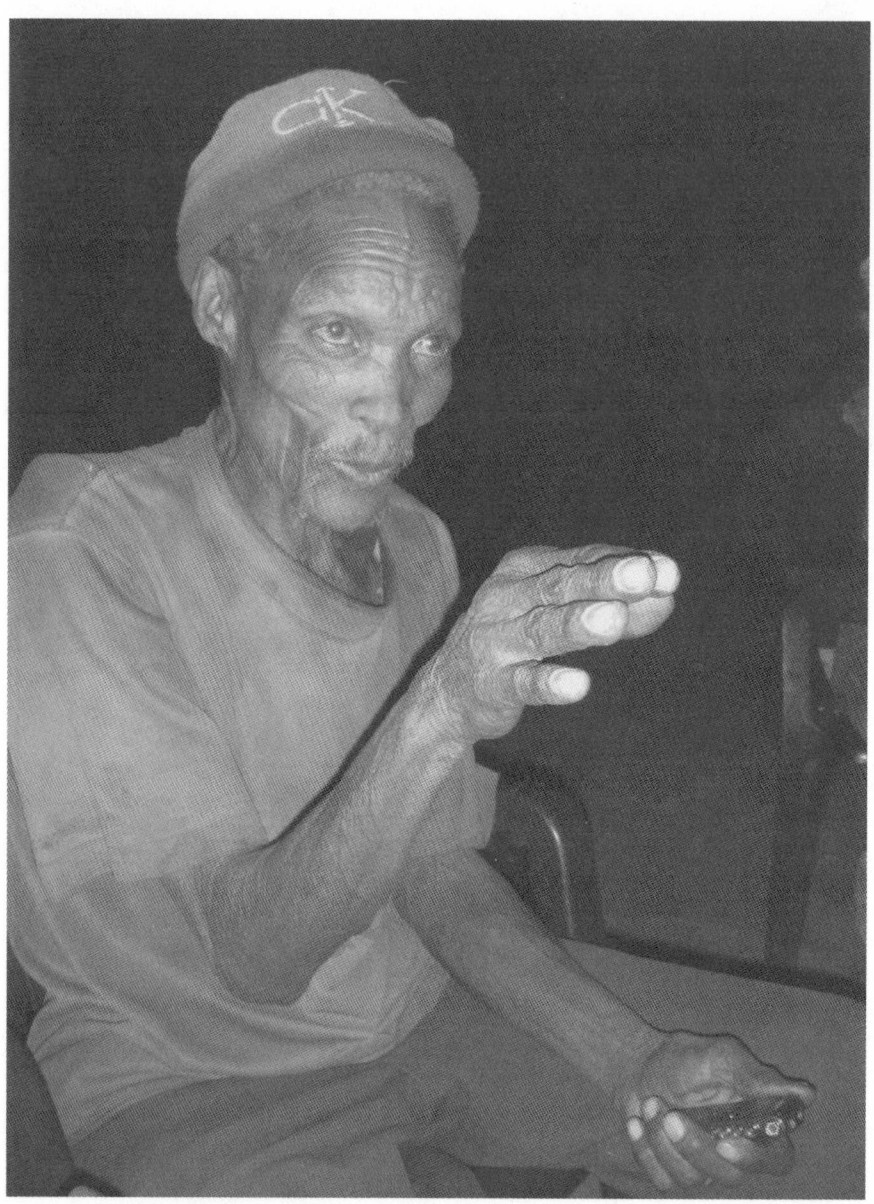

A Bushman storyteller. Dr. Polly Wiessner believes that studying their interactions around the fire can teach us much about the fireside behaviors of other hunter-gatherers, such as *Homo erectus* bands. *Image courtesy of Polly Wiessner.*

But discussions were different around the fire at night than they were during the day, Wiessner found. While some percentage of the Bushmen's nighttime conversations concerned gossip and criticism, and some were about economic matters—issues that related directly to the sustainability of the tribe and the fiscal health of the speakers' families—fully 81 percent of their evening fireside conversations dealt with *stories*, with myth, with an oral literature that serves many purposes, both instructive and diverting. Wiessner believes that stories around the evening fire, including ceremonies and celebrations, may have influenced our cognitive capacity—our ability, as she says, "to form these imagined communities, whether it's our social networks, all of our relatives on Earth, or communities that link us to the spirit world." (Of course, we may now be negating some of the cognitive and social good that lighting has produced. These days, she notes, "work spills into the night. We now sit on laptops in our homes. When you are able to work at night, you suddenly have a conflict: 'I have only fifteen minutes to tell my kids a bedtime story. I don't have time to sit around and talk.'" At first, artificial light turned unproductive time into potentially more productive—or at least, more social—time. Now, though, some of that same technology may have instead turned that potential social time into yet more work time.)

On the downside . . .

The Destructive Power of Fire

The downside of fire is so obvious that one would think it barely deserves mention: Fire can hurt. Fire can destroy. Fire can kill. Although it can be benign, it is an elemental force that can wipe out people, forests, buildings, even entire cities. (And we pay dearly for that damage. Some estimates—based largely on insurance claims—put the direct and indirect costs of fire at well over $12 billion per year in the United States alone.)

When we think of fire and the burns it can cause, we tend to stop there, knowing just how painful those burns can be. But burns actually present *two* main risks to the human body. First is the (literally) painfully obvious one we noted: Burns hurt terribly. But that raises a further question: Why *are* burns so painful, anyway?

"Burns are commonly thought of as injury to the skin from a heat source," says physician scientist Daniel Schneider of the University of Michigan Health System. "But they're more technically and broadly defined as an injury to skin and tissue cells from *any* external exposure." So, says Dr. Schneider, burn-like injuries can also result from cold, electricity, radiation, and chemicals. And if the skin is sufficiently burned, cell proteins and lipid membranes start to degrade, and then your cells lose integrity and function and die. But that's not the end of it: "When cells die," says Schneider, "they release intracellular contents that, in sufficient quantities, can be damaging to surrounding tissues."

So, why do burns hurt so much?

"What makes burns extremely painful and difficult to manage are phenomena called *hyperalgesia* and *allodynia*," says Schneider. (*Hyperalgesia* means increased pain sensation and *allodynia* is pain elicited by a stimulus that does not normally cause pain.)

According to Dr. Schneider: "The best way to describe burn pain is by relating it to a sunburn. Allodynia . . . would be the sort of pain experienced when putting on a shirt over sunburned skin. Hyperalgesia would be what one might feel if hot water were poured on a sunburn. Think of burn pain as a severe sunburn that not only burns the surface of your skin, but deep into your tissue. Burns directly and indirectly injure nerves in different layers of the skin, which is the postulated mechanism of hyperalgesia and allodynia."

And just in case the burn victim is not miserable enough, there's more. The damaged nerves Schneider mentions will transmit pain signals to the brain *even in the absence of any painful stimulus to the surrounding skin*. (This is termed *neuropathic pain*, and it's the same nasty trick the human body sometimes plays on diabetics and on amputees who suffer from phantom pains in a limb that is no longer even there.)

"Finally," notes Dr. Schneider, "what makes burn injuries extremely hard to manage is that in certain cases, commonly used pain medications such as morphine will actually *enhance* hyperalgesia rather than dampen it."

In other words, there's just nothing good to say about a burn, period.

But there's more. In addition to being painful, burns can lead to infection, and worse, because they generally result in the destruction of some portion of the one organ that stands between you and the nasty bugs of the outside world: your skin.

"Your skin is the first line of defense from the outside world," says Dr. Schneider. "Many bacteria and other organisms coexist on the surface of your skin, most of them beneficial. However, when the integrity of the

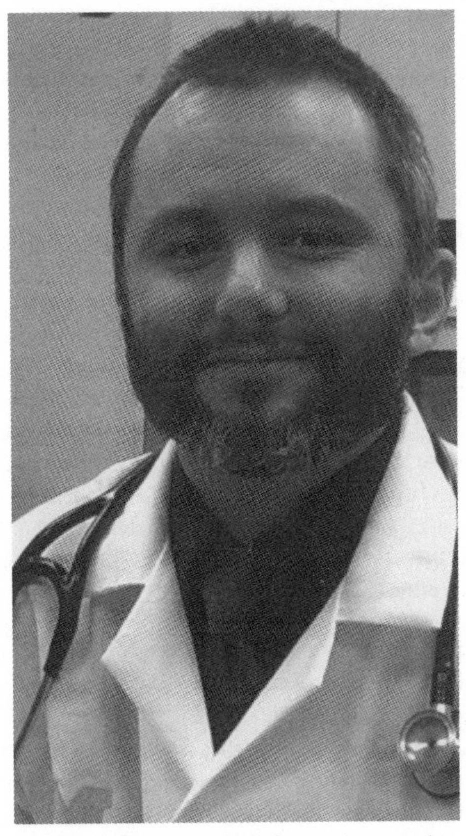

Dan Schneider is a physician and researcher at the University of Michigan Health System. *Image courtesy of Dan Schneider.*

skin is disrupted, the protective barrier is lost and there is risk for infection. It's common knowledge that many cuts and scrapes carry risk for infection, so it's not difficult to imagine that the risk increases dramatically when a significant portion of the skin is compromised. Second, the skin functions to regulate body temperature and prevent fluid loss from evaporation. People with large surface-area burns have to have their body temperatures externally controlled and require large amounts of fluid supplementation due to loss through damaged skin."

All in all, fire is not something you want to treat lightly. According to Dr. Schneider, in the United States, some 500,000 people each year are treated in emergency rooms or admitted to hospitals for burn treatments. About 3,400 people in the United States die from burns or complications of burns.

In a way, fire is like the sea: It can be beneficial, useful, lifesaving, and even beautiful; but it is not to be trifled with.

From our perspective, the interesting thing about fire is that its downside has to do with exactly the topic we've been discussing: *control.* The nature of fire, as any firefighter can tell you, is that it seeks to consume. Its supposed control is a very temporary, ephemeral thing, often illusory; all it takes is a puff of wind or a moment's distraction, and the fire that had previously been "under control" has now jumped the road or the river or the firebreak and has begun to consume homes, animals, and perhaps people. Your backyard trash fire or barbecue or bonfire is completely under control—until suddenly it's not.

All of which is similar to any other technology, of course. All technology, even those tools designed to help, build, or create, can instead be used to inflict pain, to cause distress, to destroy.

Take computers. Although certainly useful, they can be—and are now being—used to attack individuals and the infrastructures of businesses, or even of countries. A particularly virulent type of virus, such as Stuxnet, can actually cause *physical* destruction. These are effective cyberwar weapons that, in addition to damaging computers or destroying data, can actually damage or destroy physical plants. We're speaking here of nothing less than weaponized software. (Stuxnet was designed—possibly by US and/or Israeli actors—to destroy Iranian equipment suspected of being used to create nuclear weapons. More recently, an unknown actor hit a German steel mill and created massive damage to the system. In both cases, software destroyed hardware.)

The difference is that, normally, when technologies are misused, they don't take off on their own; in general, no one who used a lathe to create a weapon came back later to discover that the lathe had started itself up in the middle of the night so that it could make still more weapons. If you use a knife to attack someone, the knife does not go off by itself later to attack more people.

But fire does *exactly* that. It's why the thought of a "wildfire" is so fearsome. It is a technology that, when it goes out of control, can *stay* out of control, feeding itself and growing ever more powerful. It will burn until something stops it or until it runs out of things to destroy. (Interestingly, the virus/wildfire analogy is particularly apt. A virus *can* start running wild, infecting machines at which it was never targeted via vectors that were never anticipated. And, as one *Wired* magazine writer has commented, the real worry is not so much with well-designed, sophisticated cyberwar weapons created by nation-states, but with those that are *not* as well designed, perhaps created by terrorists or activists; those are the cyber "wildfires" waiting to happen, simply because not as much thought or effort or care went into their design.)

Tools Build Tools

As Duke University's Henry Petroski has famously said, "Tools build tools." That is, when one innovates or creates, when one builds something new or uses something old in a new way, it's difficult to see where that use will lead. When a new tool appears, it is first used for the purpose for which it was intended. But somewhere along the way, an especially bright person or group will take that tool and use it in a new way, or use it to fashion still other tools. And with *those* tools, they will build or accomplish something the original innovator could never have imagined.

Anthropologists refer to these as *positive feedback loops*.

In *How We Got to Now*, Steven Johnson lays out a fascinating series of those loops, some of them so extended and seemingly so disconnected that one would never see the associations among them without Johnson's expert (and enjoyable) guidance. And yet, they are *very* connected.

For instance, he traces the development of the printing press and explains how it caused an explosion of inventiveness not only among printers and writers and the like (which was to be expected), but also among glass and spectacle makers who now found themselves seeking optometric solutions for a seriously myopic population that had never realized its eyesight was deficient until, thanks to Gutenberg, so many people found themselves attempting to read the books, journals, and broadsheets that were suddenly available. But the loop doesn't end there. The explosion of lens crafters and experimenters ultimately resulted in the invention of both the microscope and the telescope; the former then led to germ theory and other medical discoveries, while the latter helped Galileo and others challenge the Aristotelian notion that the heavens revolved around the Earth. In a way, the invention of the printing press may have led to antibiotics and to space travel.

Closer to our time, we can envision the fathers of the Internet, men such as Vint Cerf, Bob Kahn, Donald Davies, and others, laying the groundwork for what they saw as a series of interconnected communicative nodes to be used by the military and then, a bit later, by university researchers. Could they possibly have foreseen Tim Berners-Lee's invention a few years afterward of hypertext and of a hypertext-enabled browser that would make possible the World Wide Web? And could any of them at that time have predicted the economic and social impact a maturing Internet might eventually have—how it would, for better or worse, eventually invade almost every aspect of our lives? Could Vint Cerf have foreseen a day when our thermostats and refrigerators and other household items would use some of the protocols and tools Cerf and his colleagues had invented to communicate with one another and to let us know that our supply of milk was running low or that the furnace needs cleaning or that the smoke alarm had just sounded? What would they have thought if they could have envisioned Facebook or Amazon or (and I'm truly sorry to say this) a web-enabled toothbrush?

Sir Tim Berners-Lee is the creator of the World Wide Web. *"Sir Tim Berners-Lee" by Paul Clarke—Own work. Licensed under Creative Commons BY-SA 4.0 via Wikimedia Commons.* http://creativecommons.org/licenses/by/4.0/.

Such positive feedback loops ensure that the tools that brought and continue to bring us the Internet build upon one another, often in ways no one could have foreseen, as has always been the case with technology.

The Three Stages of Fire Use

Those who study the history and effects of fire point out that early man most likely went through three stages of fire use.

Opportunistic use occurred when a natural fire swept through nearby brush or forests. The fire might have been started due to lightning strikes, volcanic eruption, or even spontaneous combustion. In a case like that, insects and small animals would have scurried away from the advancing fire front, many successfully. Others, though, would have been caught unaware and unable to flee. In examining the path of such a fire, early man would have encountered grubs or honey that had fallen from trees, burnt termite nests, and small animal carcasses. All of these would have been serendipitous finds for a tribe of hungry *Homo erectus* hunter-gatherers.

At some point—and this was a truly epochal leap for the species—someone (or some group) would have realized that fire could be approached, gathered, transported, and maintained, and that doing so was a worthwhile exercise. Thus would have begun the second stage, the *intentional maintenance* of fire. Those fires would have been useful in a variety of ways, some of which have already been noted: heat, light, nocturnal protection, and, of course, cooking.

But there would have been other advantages, too. Humans would have (mostly likely by accident) discovered the technique of meat drying; they would have realized that having fire gave them a competitive advantage when it came to scavenging; and they would have learned (also possibly by accident) that fire-hardened tips made for stronger spears and digging sticks. Over a longer period of time, they would also have discovered that controlled seasonal burning was beneficial to the surrounding ecosystems, the prairies, the brush, the forests.

By that point, of course, we were learning how to make fire, rather than merely to control and maintain it. And that's the third stage, the *purposeful creation* of fire, a stage that in turn required still other tools: bows, fire sticks, rocks for striking against other rocks to create a spark, various sorts of tinder that had to be gathered, stored, and protected. (It's worth noting that not every group has the ability to make fire, even today. The Andaman Islanders, for instance, are said [though this is sometimes disputed] to lack the ability to create fire, though they use and maintain found fire quite extensively and cook almost everything they eat, other than a few fruits.)

Archaeologist Sean Hess has talked about how Native Americans in the Pacific Northwest knew for thousands of years that forests sometimes needed to burn; that the burning cleared out the underbrush and promoted growth; that what looked like a destructive act was instead quite literally an act of cultivation.

Dr. Hess worked for the Colville Confederated Tribes, a tribe located in north-central Washington State, for several years, and he says that it was instructive to learn how the tribe saw things, how they had manipulated the environment in the past, and how they were now working to reintroduce some of those ancient forest management techniques.

"One of the things that the tribe was working on," says Hess, "was how to reincorporate fire into the management of their timberland. The Colville are a timber-harvesting tribe, and they knew that they needed fire to be reintroduced in order to make the forest healthy, because the forest evolved with fire. And a lot of it too was that the tribe purposely set fires in order to manage the forest to the state where it would be productive and attractive to them."

That's *My* Fire!

It's difficult to know whether fire was hoarded, jealously protected, and sometimes denied to other tribes or family groups. It seems like a tool that would offer such obvious and significant advantages that it might indeed purposely have been kept out of the hands of others, but many sources feel that this is in fact not the case—or at least, not the only way it could have happened.

Sean Hess feels that fire probably was *not* one of those technologies that was at first denied to others.

"It looks," says Dr. Hess, "like this was a time when people were much more egalitarian in their social structure, where everybody had basically the same set of skills, because everybody was expected to be productive. There was really not any kind of craft specialization. And so with that in mind, I think that starting out it probably would have been a relatively ubiquitous skill—working with fire, having control of that sort of technology. Especially if you think about a time in which population densities were much lower, and if you're out doing something with your family group and if you were dependent on a particular person to make fire and that person was off somewhere else, you'd be debilitated without having that skill set."

Hess points out that many of the technologies for making fire are in fact quite simple; they depend on items that anybody can easily find or quickly make, tools such as sticks and fire bows. That would seem to argue against fire being restricted to any one group or class of people.

And, as Johan Goudsblom notes in "The Civilizing Process and the Domestication of Fire," fire was itself a civilizing agent. It may have *demanded* more cooperation than conflict.

"It was simply impossible to keep a fire burning for long without at least some social cooperation and division of labor in order to guard and fuel it," says Goudsblom. "The effort of collecting fuel, keeping it dry, and putting it at the proper time onto the communal fire always involved some self-restraint, some discipline. There was no instinct specifically directing people to care for fire; it was a cultural mutation, requiring a civilizing process."

By and large, Dr. Twomey agrees with Hess and with Goudsblom.

"Cooperative foraging and regular tool use would have promoted group cooperation and social tolerance," says Twomey. "So the fire users may already have been more egalitarian than other primates. Within a social group, fire use would have promoted better cooperation, tolerance, and communication, all of which would have favored more egalitarian social structure."

However, he says, we can't tell for sure how widespread this cooperation was; local groups may have cooperated *or* competed with neighbors. "If a group could maintain access to fire while denying it to other groups, individuals in this group would have been at an advantage," he notes.

Overall, though, Twomey says that he leans toward the more cooperative view on this issue. "In the context of a breeding population made up of small social groups, the population as a whole would do better if people cooperated with their neighbors to ensure that everyone had access to fire."

Perhaps the debate—and it is indeed a long-running and occasionally rancorous one—boils down to an argument over whether human nature is essentially cooperative or essentially competitive. Many recent studies (Robert Augros and George Stanciu in *The New Biology: Discovering the Wisdom of Nature*, for instance) argue that while competition may be the norm in capitalistic societies, where it is in fact rewarded, that may not necessarily be so in other societies, or even in nature at large. Such conclusions are mirrored in Alfie Kohn's seven-year review of more than four hundred research studies dealing with competition and cooperation. In *No Contest: The Case Against Competition*, he concludes that people—and nature itself—are mainly cooperative, that competition is itself not necessarily an inherent and immutable part of human nature. (Richard Wrangham, mentioned earlier, may disagree with some of these conclusions, by the way. In *Demonic Males: Apes and the Origins of Human Violence*, Dr. Wrangham analyzes the sources of the dark side of our nature, and points out that male chimpanzees—long thought to be relatively gentle creatures—are, in the wild, often extremely aggressive, murdering, beating, and even raping. They not only compete, they kill. Given that these animals are our closest living relatives, perhaps this does not bode well for a view of man as innately peaceful.)

Then again, during the time of which we're speaking, it's difficult to tell what sort of societies existed and whether they would have tended to reward competition over cooperation. Perhaps the most likely conclusion is that both occurred, and that which one was in effect depended largely on how *erectus* defined "us" versus "them." As Twomey and others note, cooperation among ourselves (the "us") would have been beneficial for our group—whether that group was a family, a clan, or a tribe.

"While there would have been conflict," says Twomey, "there would have been mechanisms to mediate this, and most interactions would have been

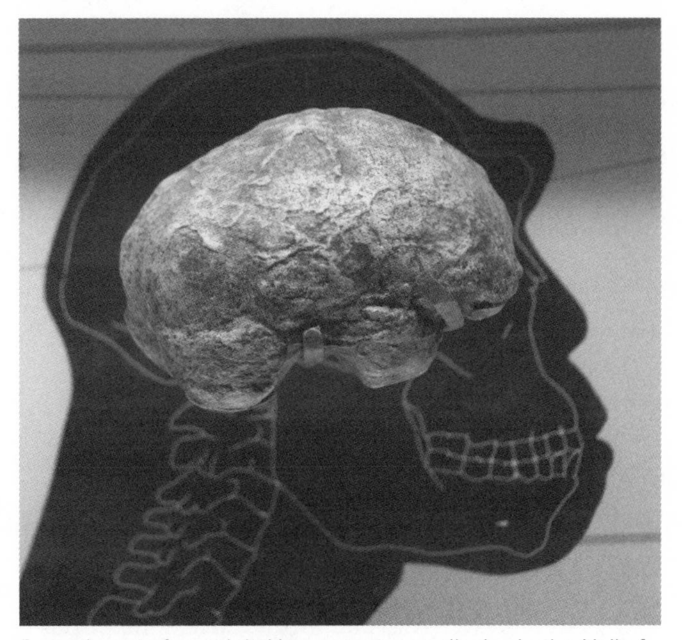

An endocast of an adult *Homo erectus* on display in the Hall of Human Origins in the Smithsonian Museum of Natural History in Washington, DC. To create an endocast, scientists fill the inside of a skull and make a model of the brain. An endocast is very useful in determining how intelligent a human ancestor might have been, and what portions of its brain were more developed. *Image licensed under Creative Commons Attribution-ShareAlike 2.0. Image created by Tim Evanson.* https://commons.wikimedia.org/wiki/File:Homo_erectus_endocast_-_Smithsonian_Museum_of_Natural_History_-_2012-05-17.jpg cooperative. It is not until more-complex social structures begin to emerge that we get war and large-scale conflict between different groups."

But there *were* forms of competition, nonetheless. In an area with increasing population, dwindling resources, and roving bands of proto-people, there would have to be.

Competition between Groups and Species

For instance, what about when one's tribe encountered another tribe? Would cooperation have ruled the day? Or would there have been enough conflict, jealousy, and suspicion to militate against whatever sharing instincts might have developed?

Another take on this debate is even broader, looking not at families, clans, or bands, but at *inter-species* competition. On this larger scale, in addition to competing with other *erectus* bands or tribes, the species that would eventually become man had first to compete with other bands of manlike apes, branches that had not evolved as far, branches that, as it turned out, would never become man. And one of the tools that *erectus* had at its disposal in this prehistoric competition was fire. The control of fire could well have been, as some sources propose, a new weapon that shifted the balance of power between our ancestors and their competitors.

When *erectus* learned to control fire, the species found itself at an evolutionary crossroads, one in which these proto-humans learned to harness a new power, one that would propel their kind forward, and one the lack of which would leave other competing species at a definite—and, as it turns out, permanent and terminal—disadvantage. Fire gave *Homo erectus* heat, light, and protection. It helped them defeat—or at least, out-evolve—competing species, and it allowed them to leave Africa and migrate across the world.

Would *erectus* have shared this new weapon with other bands? With other species?

Others have also spoken of the use of fire in this potentially competitive environment.

Zach Zorich, writing in *Archaeology* magazine, says that there was a point at which *erectus* "had achieved some cultural innovations that let them out-compete other hominin species." Among those innovations, one presumes, was the control of fire.

Similarly, Noel T. Boaz and Russell L. Ciochon, in *Dragon Bone Hill: An Ice-Age Saga of Homo Erectus*, note, "The early advantage that fire gave to *Homo erectus* was a leg up on the competition with other species—a competition that was exacerbated by climactic changes accompanying the onset of the Ice Ages."

So it seems that there may in fact have been a competitive element to the discovery and use (if not yet the creation) of fire, even if it turns out that the competition was between species as much as or more than between bands or families. In the end (well, we look at it as the beginning, actually), only one species could win out, evolve into man, and populate the world, and fire was one of the tools it used to do so.

One species did control fire, literally a world-changing technology, and that technology may not have been shared with other bands or tribes, and certainly not with other, presumably less intelligent species. The control of fire may have been restricted to *Homo erectus* simply because other species were unable to grasp either its use or its importance, but it was restricted nonetheless.

The control of fire had far-reaching effects that ultimately helped lead *erectus* to a position of primacy. As Daniel R. Headrick points out in *Technology: A World History*, "Fire allowed [*erectus*] to protect themselves from predators, to frighten animals, to warm themselves in cold weather, and to roast meat, which they needed to do because their small teeth had difficulty chewing raw meat. [Earlier species] had hunted, but only weak or small animals; otherwise, they scavenged leftovers of more powerful predators. . . . *Homo erectus*, in contrast, were big-game hunters. Working in teams, they were able to drive woolly mammoths, larger than elephants, into swamps where they could be killed with spears and stones."

And then, perhaps the successful hunters carted the meat home to be roasted in a cave in which, many thousands of years hence, a peaceful giant would live with his wife and child.

But the effects of fire do not end there, of course. As mentioned, fire may even have had a hand in helping man to achieve another important technology: the acquisition of language. In the next chapter, we'll look at that acquisition and at writing, in effect a mechanism for making language permanent and for storing and sharing information. Early writing, after all, was essentially the first form of data storage the world had ever seen.

Chapter Two

Let's Talk: Language and Writing

And the Lord said, Behold, the people is one, and they have all one language; and this they begin to do: and now nothing will be restrained from them, which they have imagined to do.

Go to, let us go down, and there confound their language, that they may not understand one another's speech.

—Genesis 11, King James Bible

Language

The truth is, we're not at all sure how our species acquired language. In fact, just *defining* language turns out not to be nearly as straightforward as one might imagine.

Consider the language of dogs. They don't have much of a vocabulary, and they're woefully ignorant of tenses. Try explaining to your dog that he "already had a cookie." Or that she can go for a walk "later." Similarly, no matter how well-trained and intelligent either of you are, your dog will probably never understand how to use a semicolon. (Then again, how many humans know how to use a semicolon?) Still, your dog is smarter than you think. There's quite a bit of cognitive processing going on in that canine brain, and a fair amount of intelligence resides there.

Take, for instance, examples recounted by canine intelligence researchers Stanley Coren (author of *How Dogs Think*) and Brian Hare and Vanessa Woods (authors of *The Genius of Dogs*). Coren notes that dogs seem to possess what psychologists call a "theory of mind." That is, unlike, say, a human toddler, *dogs are aware that their minds differ from yours*; they realize that what they see in the world is not necessarily what *you* see. This is why if you turn your back on a dog immediately after throwing a ball, when the dog brings the ball back (in itself a rather impressive cognitive feat), the dog will walk around you so that he can place the ball where you can see it. Think for a moment about what this means: The dog understands that you cannot do anything until you see the ball, and he also understands that you cannot see the ball unless it is in front of you. (And it's very important to him that you see the ball, pick it up, and throw it again and again, preferably until the dog tires out, a process that generally takes several hours. Of course, if you have more than one dog, you could be out there all night.)

Understanding that its perspective (and thus, its knowledge) differs from yours is impressive, but there's more. Consider what a trained hunting dog (and even some untrained dogs) can do.

As Hare and Woods note, dogs can understand—and even mimic—pointing behavior. A dog can show a trainer where food is hidden by running to the hiding place and looking back and forth between the trainer and the hidden food. In effect, the dog is saying, "Here! Look here, you foolish human. Just use your nose! Here's the food, right here! Wow, as dumb as you are, and you get to be in charge?!"

Even more impressively, dogs can understand a variety of gestures, and that is truly a cognitively complex undertaking. Consider a hunting dog practicing the retrieval of a dummy "duck" or other training tool. The trainer or owner throws the item far out into a field, lake, or stream, often with the dog unable to see where it landed. On command, the dog runs toward where he thinks the dummy may have landed. If he doesn't spot it, he turns back to the trainer for help. The trainer points left or right, or perhaps uses another signal to indicate that the dog is headed in the correct direction and simply needs to continue in that direction. The dog takes that in and heads off to find the dummy waterfowl. This happens day in and day out, often dozens of times during a practice session, perhaps thousands of times over the dog's lifetime. It is, quite literally, an everyday occurrence.

But consider what's really happening here. First of all, the dog knows not to take off after the "duck" until the command is given. Then he runs out to where the dummy might be. Realizing he now has no idea where the "duck" is, he looks back to the trainer for assistance.

The official portrait of President Bill Clinton's dog, Buddy, a chocolate Labrador retriever that would almost certainly have been a good hunter, had he been trained and given the opportunity to hunt. In other news, presidents' dogs get official portraits. *Public domain photo courtesy of the National Archives and Records Administration.*

At this point, the dog has just done something cognitively very impressive. The dog has indicated that he knows what he knows and, perhaps even more impressively, is aware of what he does *not* know. (Can you say that about your teenager?) In this case, he knows that he does *not* know where the decoy or dummy is. Even more impressively, he's also aware that, while he doesn't know something (the location of the training tool), his *trainer* does know. This means that the dog is aware of himself and his own knowledge; he is further aware that another being (his trainer, in this case) can know something that he himself does not. For good measure, the dog also understands his human's pointing behavior, and maps the direction in which his trainer is pointing, turning as necessary to approximate the correct route.

This is a very smart animal. Does the dog have a language he can use to express that intelligence? Most of us would say yes. A dog's body language is quite eloquent and perfectly understandable to those who are literate in "dog." We know when a dog is "play bowing" to initiate a play session, and we generally know when he is cowed, angry, happy, or excited. Of course, the dog also communicates vocally: He barks, howls, and whimpers, and surely he does these things in an attempt at communication (whether with humans, other dogs, or a possum hiding in a tree hardly matters). And the dog seems to be capable of thinking, based on the examples above and a thousand more that any dog lover would be happy to trot out. (Oddly, and somewhat disturbingly, there was a time when dogs were thought to be literally "dumb brutes," devoid of any intelligence, emotion, or feeling. Since dogs were thought incapable of self-awareness and thus unable to feel pain, the French philosopher and mathematician René Descartes and many others of the time saw nothing wrong with kicking, experimenting on, or even torturing dogs.)

Some researchers have wondered whether thought and intelligence are even *possible* without some form of language. In the case of a dog, though (and surely of many other animals), all of the prerequisites for intelligence seem to be there: The dog can think, reason, and anticipate. She can feel pain (and seemingly also happiness, love, and sorrow). She can communicate her needs and her moods. And perhaps most impressively, she can understand and respond to much of what *you* are doing, saying, and even feeling. Far from being a dumb, unfeeling brute, if these are what define intelligence, then it would appear that most dogs are more human than some humans.

Stanley Coren points out that it may be a dog's language is not based so much on words as on images and other sensory input.

"In the absence of [verbal] language," he notes, "dogs must resort to mental processes that may be similar to the sensory-based thinking that humans use as toddlers."

In other words, says Coren, dogs may substitute for words sets of images drawn from their experiences, just as we all did while we were toddlers. Even before we could speak, we were storing very important memories (potty training, who our parents were, a visit to the zoo), but since we did not yet have a language of words, we stored these as images. (This, says Coren, may be why it's difficult to recall such early memories. They surely exist, but we now use words to remember—and communicate—such events; early memories, formed before we had words, are very difficult to recall because we did not at the time have words to encode them.)

So it may be that a dog's language is mostly sensory-based. When a dog recalls his owner, he may not think "Bob" in the way we think of the word that we use to represent our dog-owning friend down the street; instead, the dog may recall a rush of images and scents that he associates with Bob. And that, augmented by the several words he does know, may be what a dog's language is like.

Mathematics as a Language

And then there is the "language" of mathematics.

When I was a high school English teacher (this was many years ago, shortly after the invention of high school), I discovered computers. Along with teaching English and journalism, I worked part-time at the local newspaper, writing features and straight news articles.

I thought it was important that my journalism students get an accurate and realistic picture of what real-world journalism was like, so I tried to model my course on practices I encountered at the paper. One of those was the use of computers to set type and, in a somewhat rudimentary fashion, to paginate the paper.

Dipping into the (fairly shallow) school journalism budget, I bought a personal computer for students to use in my class. With it, they kept track of ads, wrote drafts of stories, created artwork, and output digital files that could be sent via acoustic modem across town to the paper to be typeset.

It was, for its day, a pretty sophisticated setup for a small-town high school.

But before my students could begin using the computer, I had to master it. After all, I had to show them how to use this new machine, and that meant I had to be thoroughly knowledgeable about it, this at a time when computer skills were in short supply. The school had no IT team and no network, and the only computer-literate folks were a science teacher down

In the early 1980s, this little Commodore VIC-20 (shown with an expansion box) was the personal computer of choice for many of us. This is what was used in schools around the country and around the world, until the Apple II entered the education market in a big way. *Image courtesy of Sven Petersen, licensed under Creative Commons BY-SA 3.0 via Wikimedia Commons. See* http://creativecommons.org/licenses/by/3.0/.

the hall and one nerdy senior who probably went on to found a billion-dollar software empire. (Or at least, I hope he did.)

So I sat down at the computer with a stack of manuals and a pot of coffee, and I dug into it. Many hours later, I had discovered that I loved this machine. After thinking for many years about that first exposure to this technology, I think I can describe the feeling: Sitting in front of that now woefully laughable little computer, I felt exactly as if I had happened upon the world's biggest and best tool chest. If I could just learn to wield the tools it contained, I could build anything I could imagine. If I could think of it, I could create it. There were no limits, except my own imagination.

Imagine the feeling of power, of potential. Think of how it would have felt to have suddenly realized that you had the ability to create, to build, to innovate—and all because of this (what we now think of as exceedingly primitive) little plastic box sitting on the desk in front of me.

However, I was (and remain) an English teacher. I don't do math, at least not very well. If I add a column of numbers five times, there's every likelihood that I'll get at least five different answers. If I have to figure out a percentage, it takes me agonizing moments to recall which number to divide by which number in order to get that percentage—and in the end, I'm never sure I did it correctly.

And yet, I wanted to program. In fact, I did program, and fairly successfully, too. But it was a process fraught with clumsiness and prone to silly errors. There were many times I had to call a "real" programmer and sheepishly ask some math-related question, the answer to which any reasonably bright eighth grader surely would have known. I persevered, but it was much more difficult and time-consuming than it needed to be, and my solutions were certainly nowhere near as elegant as those devised by someone who actually had a decent math background.

So, numbers do not speak to me—or, when they do speak, they speak gibberish. But they do speak clearly to some people, and in a language all their own. And believe it or not, they sometimes say beautiful things. Listen to Josh Brown Kramer, an applied mathematician and programmer:

"I've always been drawn to eternal truths," says Dr. Brown Kramer. "Math, physics, and the more theoretical side of computer science have always had that interest for me. There is also the beauty of mathematics, which for me is realized when I come to a sudden understanding or when I realize that there is a sleek and simple solution to a problem that had seemed very hard."

Certainly, language is all about communication and problem-solving, and Brown Kramer sees in the language of mathematics both a sort of existential beauty and a real-world, pragmatic tool. Sometimes one can use something beautiful to solve problems, and sometimes a solution itself can be beautiful.

But Is Math Actually a Language?

"It is certainly fairly common to hear mathematicians call mathematics a language," notes Brown Kramer. "And it is definitely the case that there is a tight interweaving of language with mathematics. Most mathematicians would basically agree with this definition: 'Mathematics is the study of true statements.' A typical research mathematician spends a lot of his research

Dr. Josh Brown Kramer is an applied mathematician and programmer, currently working on the development of computer vision algorithms. *Image courtesy of Josh Brown Kramer.*

time making conjectures (statements that he thinks are true). A mathematician might arrive at these conjectures either by analogy to preexisting theorems (statements that are known to be true) or by running computer simulations of special cases. Another very common source of conjectures is other mathematicians. Having arrived at a conjecture, a mathematician will try to prove it. If he succeeds and it is of interest to a broader audience, he might publish it.

"There is definitely a language *of* mathematics. It is the precise language that we use to discuss ham sandwiches." (Brown Kramer is referring to what mathematicians call the Ham Sandwich Theorem. The theorem states that if you have a sandwich made of two pieces of bread and a piece of ham, then no matter how oddly shaped the pieces, and no matter where you put them, you can always make a single planar slice that cuts each of the three pieces into two equal-sized halves. Not that I know quite what a "planar cut" is. But I'm fairly familiar with ham, so let's just go with it.)

As with any other specialization, the language of mathematics includes jargon, much of it unintelligible to non-mathematicians.

"We talk about *isomorphisms*, *functors*, the *Nullstellensatz*, *polynomials*, *matrices*, *quadratic forms*, etc.," says Brown Kramer. "These are words or phrases that don't really exist outside of the realm of mathematics. There are also words that we adopt from standard language but that we use in a special way. For example, the words 'set' and 'group' mean two very precise, very different things in mathematics. The word 'normal' can mean one of several precise and very different things, depending on its context. It could mean 'perpendicular,' it could mean 'random with a Gaussian distribution,' it could mean normal as in 'normal subgroup of a group.' We're very particular about what the word 'or' means. Specifically, it is the inclusive or. So, as a mathematician I would say, 'We'll let him in the group if he can kill zombies or he can cook' without feeling compelled to add 'or both.' Another example along those lines is that the phrase 'All that glitters is not gold' has a very different meaning than 'Not all that glitters is gold.'"

There was more to my interview with Dr. Brown Kramer, but he suddenly began speaking math, and so I had no idea what he was saying.

Where Did Language Come From?

Here's the short answer: We simply don't know.

There are all sort of interesting (and sometimes fanciful) theories about the origin of language, but no proof. (One early theory posited that for every object in the world, there was one correct word to describe it. When that word was found and used, a sort of cosmic "bell" rang to indicate that this was indeed the "right" word. It will surprise no one to find that this hypothesis was known as "The Ding-Dong Theory.")

There isn't even any real agreement on the *speed* with which the acquisition of language took place. Some linguistic ability seems to have arisen slowly, evolving over time, as did the species itself. But some scientists point to the possibility of a sudden, fairly recent event that may have been triggered by a single genetic mutation. (One thing on which most scientists *do* agree is that language, like the species itself, originated in Africa.)

The lack of solid proof about the roots of language is not really very surprising. The University of Melbourne's Antonio Sagona points out that the origins of language are much more difficult to track than, say, the origins of writing (a subject on which Sagona happens to be an expert), because while we have recorded or hard evidence for writing, our evidence for language is circumstantial, namely neurological inferences drawn from patchy physical anthropological data. Most of that comes from skeletal material rather than from speech itself, of course, given that speech is invisible and ephemeral. But that material is difficult to find.

Dr. Sagona notes, "Neuroscientists tell us that for language to develop, hominids had to have developed parts of the brain, including Broca's area. And to see that, you need a near complete endocranial cast, something which is very rare indeed."

About the only thing that language experts agree on, says Sagona, is that symbolic and ritualistic behavior probably requires a fairly sophisticated communicative system. So, when we see evidence of art, of intentional burials with grave goods (i.e., items buried along with the body, such as personal possessions and supplies to ease the deceased's journey to the afterlife), etc., these most likely required language. And these are fairly recent events, say, within the last one hundred thousand years. So, however language developed, it may be of relatively recent vintage.

Regardless of how and when we acquired it, we *need* language. A human who lacked language would in fact be something less than human.

As Karen Armstrong notes in *The Bible: A Biography*, language does more than merely help us to communicate about our day-to-day needs. Humans seek meaning, stability, understanding, and pattern. Language provides that meaning and that pattern, and from it we derive the understanding we seek.

"Language plays an important part in our quest," says Armstrong. "It is not only a vital means of communication, but it helps us to articulate and clarify the incoherent turbulence of our inner world. We use words when we want to make something happen outside ourselves: we give an order or make a request and, one way or the other, everything around us changes, however infinitesimally. But when we speak we also get something back: simply putting an idea into words can give it a lustre and appeal that it did not have before. Language is mysterious. When a word is spoken, the ethereal is made flesh."

Frozen Language: The Origins of Writing

But *language* isn't enough. We need writing, too. As Antonio Sagona has noted, it would be difficult to imagine a world without the ability to record language, to record thoughts and ideas; using a flexible, communicative system of recording our thoughts is fundamental to our daily existence.

In a sense, writing is language that has been preserved, stored in the form in which it was first uttered, and thus able to be passed on from one person—and from one generation—to another, exactly as written. It is a representation of a thought or idea that has been frozen in time; the act of reading allows one to recall exactly what the writer said, right or wrong, and whether said well or poorly. Writing is visible speech.

This is important. There is much to be said for an oral tradition (and some great literature and important knowledge were created and passed down using that tradition), but accuracy is not necessarily one of its hallmarks. When we passed on information orally, words changed, myths evolved, ideas were lost (and then sometimes found again), and legends were conflated and confused. As good as the preliterate storytellers were (and they certainly had prodigious memories and a huge store of knowledge), they were nonetheless imperfect, and their stories appear to have been subject to a certain amount of selective revisionism.

In addition, oral communication is simply limited in terms of what it can accomplish.

"Writing encouraged complex thought," says Dr. Sagona. "There is a limit to what you can do with the spoken word. You cannot, for instance, deal with advanced numeracy and scientific observations, nor can you engage in complex arguments such as legal procedures—these must be written down. Over the course of a millennium or so, writing developed from an account-keeping system to a creative tool. You needed it to 'create' governments, bureaucracies, literature, and so on."

So, writing represented an incredible improvement in the transmission of ideas, of history, and also of literature. Without it, there could be very little in the way of accumulated knowledge, history, or science. Writing is fundamental to our existence; without it, few other technologies would exist.

Dr. Antonio Sagona is an archaeologist and a professor at the University of Melbourne. He specializes in archaeology of the Near East, has written several books on the subject, and is the editor of the journal *Ancient Near Eastern Studies. Image courtesy of Antonio Sagona.*

The Evolution of Writing

So where did writing come from? Ah, now here we have more evidence at our disposal than when we were speaking of the birth of language.

Our examination of the origins of writing starts in Victorian London, and ultimately includes brief contacts with such notables as Sir Winston Churchill, T. E. Lawrence, and mystery writer Agatha Christie. But let us begin with the director of a multiyear excavation in what was once ancient Sumer.

Charles Leonard Woolley was born in London in 1880. He attended "public" schools (that is, what would be termed "private" schools in the United States) and then Oxford, and from an early age showed an interest in, and an unerring instinct for understanding, excavations and antiquities. It was almost a given that Charles, later Sir Leonard, would become an archaeologist.

And what an archaeologist he was. Setting his sights on ancient Sumer, and funded by the University of Pennsylvania's Penn Museum, Woolley headed for Britain's Colonial Office to get permission to excavate in what would eventually become Iraq. The head of that office was Winston Churchill, and Churchill's assistant secretary was none other than T. E. Lawrence—Lawrence of Arabia. Lawrence worked with Woolley to help arrange the expedition. In September of 1922, Woolley and his team set out from London for Basra, a port in the south of the country. It was the historic location of ancient Sumer, and Woolley and his team were hoping to find a virtual (and possibly literal) treasure trove of artifacts.

They did. The site, which over time became many separate sites all linked to the Woolley expeditions, was so productive that Woolley and his team traveled there yearly until the 1930s, when the Great Depression put an end to the funding, and thus to the expeditions.

T. E. Lawrence (left of slab) and Charles Leonard Woolley (right) exhibit a Hittite slab at the excavation site at Carchemish. *Image in the public domain.*

On the Woolley team was Max Mallowan, a young Oxford graduate who came aboard as Wooley's assistant in 1924. At Ur, Mallowan (later Sir Max Mallowan) would meet a well-known friend of Woolley's wife: Agatha Christie. The mystery writer married Mallowan in 1930, and would in fact eventually write *Murder in Mesopotamia* (a melodramatic potboiler in which the murder victim, a putative widow, turns out to have unknowingly remarried her *first* husband, not realizing that he had not died in a train crash after all), basing it on her travels there with her husband. (They were, by all accounts, a devoted pair. From the time she married Mallowan, the two of them never traveled anywhere the other could not work. Christie traveled the world with Mallowan, setting up her little "office" near dozens of digs, where she wrote mysteries while her husband excavated ancient sites.)

During the Ur excavations, Woolley uncovered thousands upon thousands of important artifacts, and endeavored to use them to construct an amazingly complete (and occasionally fanciful) history of the region. Woolley excavated what turned out to be a royal cemetery, uncovering several royal tombs and recovering jewelry, tablets, and precious metals, revealing evidence that Sumerian court attendants were killed and buried alongside the kings and queens that they had served.

All of these findings were romanticized by the press, and the popularity of Woolley's expeditions among the lay public reflected the 1920s "golden age" of archaeology that gave us the discovery of Tutankhamun's tomb. (And the "curse" that supposedly resulted in multiple deaths after archaeologist Howard Carter and his team entered the tomb in 1923.) As unlikely as it may seem to us today, Woolley and his team of sunburned academics were heroes back in England and the United States.

(On a side note, Woolley was occasionally given to extravagant interpretations of his discoveries. For instance, he was positive that he had found proof of Noah's flood in the strata of his excavations. But since other architects had—and have since—found similar "proof" in multiple places and along multiple timelines, it is likely that we're seeing evidence of several floods—deluges certainly, but not necessarily sent by God to cleanse the Earth. As it happens, in Ur, where Woolley was excavating, violent, sudden floods are not at all uncommon; Mallowan himself described a torrent that began with the workers standing on perfectly dry land just outside the house where the team lived and ended, only minutes later, with them in waist-deep water.)

Woolley was lionized by the press and by his teammates. Eager to engage the public and a gifted writer, he was skilled at conveying history in the guise of thrilling stories that could be understood and enjoyed by lay readers. Returning home, he wrote several wildly popular books recounting the Ur expeditions and explaining the team's finds to an adoring public, intensely interested in hearing about ancient Mesopotamia.

Mallowan said about Woolley, "Woolley's observations missed nothing and his imagination grasped everything."

But Mallowan was wrong. Woolley *had* missed something. Something that would turn out to be very important.

"Miscellaneous Clay Objects"

Archaeologists in the field are very meticulous about their work. They have to be; if they were not, the entire enterprise would quickly descend into chaos. No one would know what artifact was found where or at what level of the dig, or if a given item was in fact a real artifact or just some piece of rubble that had somehow made its way into the pile. No one would be able to assemble a historical narrative that would square with what was discovered at the site. It would be a terrific waste of time, money, and resources, and a terrible squandering of the historical record.

Field workers thus labor diligently to ensure that the record is accurate, and one of the things they do is keep very careful, very complete notes, known as *field notes* or *site notes*. Everything the team uncovers is entered into what amounts to a manifest, along with information about exactly where each item was found and comments about what the find may mean. At a site such as Ur, this could mean recording the discovery of many thousands of items: pieces of jewelry, tablets, tools, pottery, decorative items, and much more. Every one of the items had to be recorded in the site notes.

But what about little bits and pieces of what appear to be trash? Or fragments that are so small that no meaning can be attached to them? What do archaeologists do about little miscellaneous bits of rock or clay that could be . . . well, *anything*—or nothing? They could literally be garbage from a recently abandoned dwelling, or they could be some nub that broke off of a larger ancient piece and which was therefore thrown away—in other words, garbage again, but age-old garbage this time.

Archaeologists do find such things, and they don't always know what to make of them. Woolley, for instance, found well over one hundred bits of formed and fired clay or stone. They meant nothing to him, and like other archaeologists of the time, he simply recorded them as "miscellaneous clay objects." To him, they were insignificant debris. (Some archaeologists actually discarded them, assuming that they were meaningless bits of rubble. Some surmised that perhaps the little conical or disc-shaped items were gaming pieces. That actually made a certain amount of sense, because the area was in fact the birthplace of what we would call board games, and game pieces of that era ranged from pedestrian and utilitarian to tiny works of art, often presented to rulers as symbols of the givers' fealty and devotion. In fact, in August of 2013, researchers from Ege University discovered five-thousand-year-old gaming tokens in what is now Turkey.) In the end, many of these items were either discarded as worthless or lumped together as bits of miscellaneous rubble.

Woolley and the others were wrong. The small clay items were not garbage, they were not rubble, and they were not gaming pieces. They in fact signaled the beginnings of the evolution of writing.

Accounting as Writing

The import of these small clay objects remained undiscovered until two researchers at the Louvre decided to examine them more closely. Archaeologist Pierre Amiet, and later, his student, Denise Schmandt-Besserat, thought that these little bits of fired clay were in fact tokens meant to help people keep track of quantities of goods. Sheep, oil, sheaves of wheat, or almost anything could be tallied by using tokens of different sizes and shapes to represent various quantities of various goods. The tokens were then encased in hollow clay balls, which we've come to call "envelopes," as a way of gathering and storing multiple tallies while avoiding loss or tampering. The envelopes themselves were inscribed with markings meant to convey the contents of the clay balls.

Dr. Schmandt-Besserat, in *Before Writing, Vol. 2*, painstakingly cataloged over five thousand of these tokens from digs all over the Middle East, over one hundred of them from the Woolley Ur expeditions.

These bits of clay are called *tokens* for a very simple reason: They stood for something. They represented both quantities (twelve jars of oil, for instance) and concepts (e.g., the idea of sets of items), and the ability to indicate these in a semipermanent fashion heralded the birth of writing.

From Accounting to Writing

Our hunter-gatherer forebears had no need to keep track of things; after all, they *had* very few of them. And, being foragers, they didn't stay in one place long enough to accumulate many things. One might have a club or a spear, a flat rock on which to grind wild grains, some small stone tools, and perhaps a woven basket or bag in which to keep these few items. Generations of proto-humans—and later, true humans—got by with very few possessions. It simply doesn't take much to wander the plains, looking for food and shelter. (And, anyway, the more possessions you had, the more difficult it would be to pick up and move when the food ran out, the water dried up, or you were simply chased away by predators—human or otherwise—larger and better armed than yourself.)

And then, about ten thousand to twelve thousand years ago, things began to change.

For centuries, people had been haphazardly tending undomesticated animals, sometimes herding them as they foraged. Similarly, they had been gathering plants, sometimes carrying them with them on their travels. (And, one assumes, occasionally planting seeds, perhaps in the hopes that edible plants would flower before it was time to move on.)

But over time, the combination of natural selection and occasional mutation produced plants and animals that were increasingly reliant on people. And the people in turn discovered that they could stay in one place—sometimes for a season or two, sometimes permanently—while raising these plants and animals. If nothing else, having a store of domesticated plants and animals helped them through the lean times and allowed them to supplement the game they hunted and the plant foods they gathered.

Somewhere along the way, though, the situation reversed: People discovered that raising domesticated plants and animals for food was more reliable than hoping to find food in the wild. (It's also tempting to assume that perhaps agriculture was also less effort than foraging, but almost any farmer working today would be glad to disabuse you of that notion.) Now, with the establishment of what would eventually become farms, ranches, and orchards, hunting and gathering had for most people become ways to supplement the food people grew and raised, rather than the other way around.

With the rise of farms came permanent settlements, of course. And settlements became towns and then cities, and cities (and entire regions) were linked by trade. Suddenly (at least, "suddenly" in terms of the immense span of history), people were trading goods: wheat for oil, cattle for wheat, leather for foodstuffs, and sheep for clay containers in which to store some of this food. Some people produced surpluses large enough to take into the center of what had become a town, and sell their surplus at a market or bazaar that had sprung up in the town center.

Now people *owned* things. Land. Homes. Sheep. Oxen. Bags of grain, and grindstones with which to grind that grain into flour. There was now a need to record quantities of items sold or lost or borrowed. I could trade vessels of oil to a farmer for bags of wheat, but someone had to keep track of that oil and of that wheat. In fact, I could even borrow today against a quantity of wheat I would be expected to deliver in the future—but someone had to record and remember who owed what to whom.

Accounting, you see, had suddenly become necessary. And it had become necessary solely because of the new agricultural economy. And that record-keeping, the accounting of barrels of oil and sheaves of wheat and all the rest, is what led to writing.

In an economy of producers and suppliers, lenders and borrowers, one must have a way to keep records. Thus, the invention of tokens: small discs and spirals of fired clay that at first stood for various quantities and, eventually, for the ideas that those quantities represented.

From a technological perspective, says Dr. Sagona, writing was a tool to record transactions, a way to control some of the communications in what had become large cities. The population in places such as Uruk, in southern Mesopotamia, had reached forty thousand or so inhabitants, and at that point it was no longer possible for everyone to remember important details about contracts, agreements, and deliveries of goods. Villagers who lived in settlements of two hundred to three hundred inhabitants could still rely on memory, but urban areas with centralized institutions required various technologies to maintain control and order.

One of those technologies was writing.

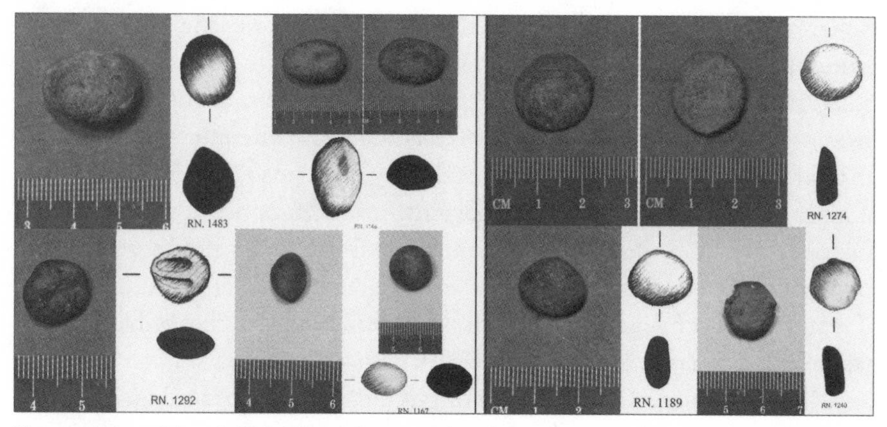

These clay ovoid and disk tokens from ancient Iran were used in a farming community to tally livestock and grains. The use of such tokens not only led to writing, but also affected social structure, politics, and overall cognitive skills. *Image courtesy of Antiquity, Department of Archaeology, Durham University, Durham, UK. Fazeli Nashli, H. and N. Moghimi. 2013.* "Counting objects: new evidence from Tepe Zagheh, Qazvin plain, Iran." *Antiquity 87(336 Project Gallery). Available at: http://antiquity.ac.uk/projgall/nashli336/.*

"Initially," says Dr. Sagona, "writing was developed as a means of record keeping. At first writing was produced as pictographs—images and symbols that closely related to the item depicted."

This is where the clay tokens came in. The token system, says Dr. Schmandt-Besserat, was the earliest system of signs used for transmitting information from one person to another and from one community to another. At first, each token represented one symbolic concept: Cones and spheres represented measures of grain, cylindrical or disc-shaped tokens represented a unit of animals, and other shapes represented other types of products.

But clay tokens, after all, are small and fragile. They can get lost, broken, or even stolen. And groups of tokens can be miscounted or tampered with. The solution was to place groups of tokens inside of hollow clay balls, the "envelopes" mentioned earlier, as a way of gathering and storing multiple tallies. The envelopes themselves were inscribed with markings meant to convey to the reader (and perhaps this is the very first time that the term *reader* can accurately be used) the contents of the clay balls. One can thus imagine an envelope within which a number of tokens are rattling about, and inscribed on the clay ball are figures noting that this particular envelope contains a certain number of tokens denoting the three ewes and a lamb that are owed to Gitlam, now in possession of the tokens, and for which the tokens are meant to stand as credit, from Penzer, who borrowed from Gitlam. (And if Mesopotamian names sound a bit odd to our ears, we can always substitute "Kyle" and "Steve" instead.)

It is but a small step from marking symbols on the clay balls that held the tokens to simply doing away with the tokens altogether, flattening the clay balls into tablets, and then marking the tallies on the tablets themselves. Eventually, cuneiform script, the first true writing system, emerged from what had essentially been a counting device.

So, a generalized token system had evolved into mnemonic devices used with what can only be described as accounting tablets; we had created a sort of a Mesopotamian VisiCalc.

When Writing Became Phonetic

At first, writing was essentially mnemonic: Images on the envelope represented items—a ewe, a pair of oxen, a pitcher of oil. But then came a crucial breakthrough. "There came a point," says Sagona, "when images stopped representing the items they looked like and instead began to stand for the *sound* of the words used to describe the thing being depicted. This is when writing became phonetic."

And when writing became phonetic, when it could reproduce the sounds of human speech—and, therefore, speech itself—it could develop beyond accounting to help create literature, history, and poetry. It could represent and communicate the thoughts of the common people and the will of rulers, and could thus encompass law and philosophy. With writing, governments could exist.

In the end, it was agriculture, commerce, and politics that led to writing. As with many technologies, evolving political institutions and new types of commerce created a situation that *required* a new technology—writing—and thus, it was invented.

On the downside . . .

The Pre-Luddites. And the Post-Luddites.

It's difficult for us to conceive of a downside to writing. We could argue that, as with other technologies, writing could be misused—manipulated to spread messages of hate or misinformation, or to incite unrest. (Some of the very things we *do* sometimes view as the downside of the Internet and social media, in fact.) But writing is so basic and so fundamental to our existence that in general, we simply accept it as something . . . well, *necessary*. Something so basic that we cannot imagine life without it.

But not everyone was happy with the invention of writing. Some felt that it enabled people to *appear* knowledgeable when in fact they were not. It was, said some elders, a crutch, an "easy way out." (Why, it'd be almost like allowing students to use calculators and computers in math classes, one supposes.)

Socrates (via Plato's *Phaedrus*) recounts the myth in which the Egyptian god Thoth (or Theuth), said to have been the inventor of writing, seeks the king of Egypt's blessing and praise. But instead of praising him, the king says, "You have invented an elixir not of memory, but of reminding; and you offer your pupils the appearance of wisdom, not true wisdom, for they will read many things without instruction and will therefore seem to know many things, when they are for the most part ignorant."

And as Cal State University's Susan Dobra has noted, "Not only through the mouthpiece of Socrates in the *Phaedrus*, but from his own stylus in *The Seventh Letter*, Plato denies the legitimacy of the written word as capable of conveying knowledge in any truly significant way."

Ironically, Plato himself communicates his ideas about the problem with writing *in writing*. Why? Well, as Dobra points out, "If he hadn't, few of his ideas would have survived much beyond his own lifetime, their being too complex and detailed for any hope of accurate oral transmission."

The complaint, echoing from a few thousand years ago, sounds familiar. We're constantly bemoaning the negative effects of technology on young people, and especially on their facility with language.

Texting is one example. "Students are so used to texting," we're told, "that they've forgotten—or perhaps never learned—how to write." They use weird abbreviations, acronyms, and emoticons, and they write in brief, telegraphic bursts that bear little resemblance to the kind of sophisticated language we like to see people use when writing. Technology has once again acted to the detriment of the language, or so many have come to believe.

Dr. John McWhorter, a linguistics professor at Columbia University, begs to differ. He says that the reason we're fixated on the idea that texting is harmful to the language—and to the skills of young people who speak and write the language—is that we're looking at texting in entirely the wrong way. Texting is not writing at all, he says. It is instead "fingered speech."

"Linguists have actually shown that when we're speaking casually in an unmonitored way, we tend to speak in word packets of maybe seven to ten words," Dr.

Linguist John McWhorter argues that texting is *not* writing and that it is not damaging the language. Rather, he feels that texting is a form of what he calls "fingered speech," and ought to be considered as something very different than writing. *Image courtesy of Dr. John McWhorter.*

McWhorter noted in a 2014 TED talk. "You'll notice this if you ever have occasion to record yourself or a group of people talking. That's what speech is like. Speech is much looser. It's much more telegraphic. It's much less reflective—very different from writing. So we naturally tend to think, because we see language written so often, that that's what language is, but actually what language is, is speech. They are two things."

Texting, says Dr. McWhorter, is much more akin to speech than to writing. The former tends to be brief, casual, full of linguistic shortcuts, and augmented by body language: a wave of the hand, a smirk, or a raised eyebrow can add much to—or even completely change—the meaning of what someone is saying. When we compare writing to texting, says McWhorter, what we're looking at are two very different things. The former is organized and formal discourse, while the latter is simply a much more casual form of communication. In effect, texting is speech—what McWhorter calls "fingered speech."

Students' (and to be truthful, often their parents') preoccupation with technology may be doing any number of things—some of them surely bad—but destroying their ability to write is probably not one of them. (If that *is* happening, then perhaps it's time to look elsewhere for the cause. Grade inflation? Lax parenting? Laziness? Well-meaning but overbroad pronouncements from educational "experts" who've not been in a classroom for twenty years or more—or ever? Not that I'm bitter.)

The Literate Elite

Early writing was complicated. Cuneiform script eventually used a system of from five hundred to six hundred discrete signs. Learning those symbols, and how to use them, was a demanding, esoteric skill, one that required years of study to master, and that sort of study was expensive. In the end, only the very wealthy could afford to spend the money that was required to make their children masters of this almost priest-like skill. Therefore, writing became a way to distinguish the literate elite from the rest of the population.

"Most people who lived in the heartland of writing were illiterate," says Antonio Sagona, "and the ability to read and write was restricted to a privileged few—the scribes."

In some instances, the scribes knew several languages, and were also quite cocky about their skills. Mesopotamia had very good schools, but not everyone had access to them. The interesting thing is that even though most people could not read or write, writing—whether in the form of tablets, or obelisks with laws engraved on them—was seen as powerful, almost as something spiritual. "Writing," says Sagona, "symbolized power, respect, and complexity." Like reading, writing was a key to understanding and a gateway to power and influence.

This is the point at which what amounted to an intellectual caste was created, possibly for the first—but certainly not for the last—time. It resulted in a situation in which only the already wealthy (read: powerful) were able to read and write. Now, keep in mind that much of what they were writing comprised laws, bills, deeds, and pronouncements from local, regional, and—to the extent that these existed at the time—national governments. Thus, people with wealth and power were able to help create instruments of power, while the people who lacked that technology, were, by and large, excluded from the exercise of power. (So, really, not much has changed, except that now many more of us less-powerful people can also read and write.)

It is likely that the powerful, those who had controlled writing up to that point, looked askance upon the idea that eventually the skill would be passed on to commoners. Just as slave owners in nineteenth-century America prohibited slaves from learning to read and write—and for essentially the same reasons—those who wielded power in ancient Sumer cannot have been pleased when this particular playing field began to be leveled.

Writing is thus the poster child for our theme: the idea that technologies tend to start out in the hands of the powerful and only later trickle down to the rest of us. Once this technology spread and became relatively commonplace, anyone could use writing to his advantage.

Of course, whatever writing *was* produced during this time—and for many, many years to come—was produced very slowly, and it remained in limited circulation. Its effects were surely revolutionary, but restricted. In the fifteenth century, this changed rather abruptly. A knowledge revolution began in a small city in what would eventually be the Alsace region of France, near the German border. We are still seeing the impact of the revolution, and its cause will be our next topic: the coming of true books and, eventually, printing via movable type.

Chapter Three

A Communication Revolution: Books and Printing

Unforeseen consequences stand in the way of all those who think they see clearly the direction in which a new technology will take us. Not even those who invent a technology can be assumed to be reliable prophets, as Thamus warned.

—Neil Postman, in *Technopoly: The Surrender of Culture to Technology*

The Successful Failure

It was just another start-up. Understaffed. Low on cash. Crammed into workspaces that were cramped and uncomfortable. Trying desperately to hold off the slavering creditors long enough to get the new product to market. Working day and night to create something that might astound the world—or something that might simply suck up all the resources they could muster and then never show a penny of profit. The enterprise lived on borrowed money, with stakeholders clamoring for a product they could sell to recoup their investments. It was a scene familiar to anyone who has been part of a start-up.

But this little company was different. Situated in Mainz, Germany, near the west bank of the Rhine (or Rhein), it was a ragtag outfit run by a former goldsmith turned printer named Johann Gutenberg. (His name was also rendered as Johannes or Johanne, among several other such variations. In English, we would call him *John*.) He was attempting to do what no one else had done before him: Gutenberg wished to create a press that used movable type, so that the individual letters that formed a page could be reused to print a different page—or a thousand different pages. If he succeeded, he might change the world. If he failed, well . . . his could be an ignominious end; he might end up behind bars, given that debtors' prisons in Germany were not eliminated until the mid-nineteenth century.

In retrospect, Gutenberg's idea was simple—as some of the most important and most powerful ideas often are. Instead of carving letters and entire lines of text in a block of wood or other medium and then inking that block and pressing it on the paper or parchment (an impressive artisanal skill still very much in demand in certain quarters), Gutenberg proposed a new approach. He wished instead to create individual letters in metal (he tried using wood, but that didn't work particularly well), and then to encase the metal letters in a frame called a *chase*. He would then place the chase in a form, ink that, and impress upon the paper the image of the pieces of type locked in the chase. If it worked, this would constitute an astounding

An image familiar to anyone who took print shop in school before about 1980 or so, this photo shows movable metal type and a composing stick not all that different than what Gutenberg or his fellows might have used.

advance in the printing of books, pamphlets, newspapers, or anything else. What had previously taken dozens of craftsmen months or even years to do could now be accomplished in days. The price of books could plummet; perhaps people previously unable to afford printed books could now purchase and use them in their businesses, possibly even in their homes.

This was one of those brilliant ideas that actually required very little in the way of true invention. Innovation, yes, and also intelligence, ingenuity, and persistence—lots of persistence. But most of what Gutenberg used on the path to his monumental achievement already existed in one form or another.

The Process

Printing, as Gutenberg envisioned it, required three critical things: movable metal type, an ink that would adhere to the type, and the press itself. By the fifteenth century, the press already existed. (Many were, as Gutenberg's would be, adapted from wine presses of the day.) Ink certainly existed, though Gutenberg would in the end need to reformulate his so that it would adhere to the metal type he would eventually create.

This is a species of problem familiar to high-tech innovators: A brilliant researcher creates a new technology, but then discovers that additional new or enhanced tools are required in order for the new technology to work effectively. With automobiles, for instance, the creation of finely calibrated, reliable engines ultimately depended on efficient fuel delivery, something that would only be possible once someone managed to invent the carburetor. (Karl Benz received a patent on such a device in 1886, though his invention may have been preceded by that of two Italians, Luigi De Cristoforis in 1876, and Enrico Bernardi in 1882, and by an Austrian, Siegfried Marcus, even earlier.)

With the 3D printer discussed in chapter 8, developer Chuck Hull had to create new software and hardware in order for the new technology to work reliably, consistently, and affordably. Many an inventor has spent much of his time jumping down such rabbit holes, trying to resolve issues that were created by the very technology on which he was working. Some never do emerge from those holes, and the once-promising new technology founders.

Most of what Gutenberg needed already existed in some form. The real trick here was in casting the type: creating individual letters that could be assembled, inked, used to create multiple impressions, and then reused in different—and almost infinitely variable—combinations.

But even this process already existed, in however rudimentary a form. Goldsmiths (and keep in mind that Gutenberg was one) had long used carved punches to ornament the spines of bound volumes; similarly, artisans working with pewter had for many years been using brass die-stamps to mark and decorate their work; and craftsmen in the employ of various government mints had for centuries known how to create coins by casting metal in a mold. Gutenberg's approach required that a skilled artisan create a punch by carving a letter backward in the tip of a metal rod. (No mean feat, that.) The punch was then driven into a softer metal called a *matrix*, and the matrix was placed in a mold, which was filled with hot lead to create the individual pieces of type. The punch and the matrix were reusable, so that

This printing press, made in the early 1900s and on display in Haifa, Israel, works not all that differently than Gutenberg's would have. *Image used under the Creative Commons Attribution-Share Alike 3.0 Unported license.*

once the font itself was created, making pieces of type to fill the chase was quite an efficient operation. (New research suggests that Gutenberg may in fact have used molds made of sand; this would have been somewhat less efficient, but the process itself wouldn't have changed much.)

Thus, Gutenberg's genius lay not so much in creating new things, but—as is often the case with innovators—in utilizing existing technologies in a very new way. We'll return to Gutenberg shortly, but first a few words about what led to the movable-type revolution: books themselves.

The Earliest Books

Possibly the impetus for printed books as we know them today came from woodcuts, or *xylography*: a process that originated in China in which an artisan carves an image (or text) into the surface of a block of wood, a process that requires an incredible talent and quite a deft touch. It was not a huge leap from carving images in relief in wood to carving letters—or entire texts—on such blocks. But these woodcuts had to place ink on *something*, first on animal skin, and then on paper.

As one author noted, "The very existence of texts and pictures printed with wood-blocks may have made the possibility of using paper for the mass-production of texts more evident, and quite probably the success of block prints and books made it possible to foresee the kind of success that a more perfect process might enjoy. In short, it is possible that the widespread use of block-books spurred Gutenberg's own initial enthusiasm."

Paper came to Europe via China and, prior to that, via the Arabs. So, along with algebra, the guitar, coffee, and the toothbrush (perhaps those latter two are connected), we have Muslim Arabia to thank for paper.

It began showing up in twelfth-century Italy as what was then considered a new form of "parchment," though more fragile than actual parchment, and easily torn. (True parchment is untanned animal skin. Though the term is often used interchangeably with *vellum*, the latter is actually made specifically of calfskin, while parchment can be made from the skin of any animal. The technical distinction is not always made these days. In any case, what we call "vellum" today is often made of a synthetic material, rather than of animal skin.)

By 1398, artisans in Paris had formed a papermakers' guild, and in 1489 King Charles VIII granted rights to those who were officially allowed to make and use paper, including the papermakers themselves, as well as scribes and binders. (Charles was known, for no reason that springs immediately to mind, as "Charles the Affable." Perhaps he was easy to get along with because he was, as one historian has said, "feeble in body and intellect." He may also have been extraordinarily tall for the time, given that he died at the age of twenty-seven after hitting his head on the top portion of a doorjamb, having reigned for only fifteen years.)

The expense of parchment may be one of the reasons that the cost of books remained so high for so long.

Dr. Stephen Buhler is the Aaron Douglas Professor of English at the University of Nebraska at Lincoln, and the author of *Shakespeare in the Cinema: Ocular Proof*. He specializes in the literary culture of Early Modern England, and often discusses with his students the effects of technology on that literature.

"Geoffrey Chaucer continued working on *The Canterbury Tales* until his death in 1400," says Buhler, citing one example. "He wrote for a manuscript culture: every book in existence had to be copied by hand on parchment—animal skin. That combines a labor-intensive product with expensive materials, so books were relatively rare and ownership limited to few."

But, as we'll see, books did not *remain* rare, nor their ownership limited.

The book itself can be said to have originated almost as soon as writing itself began (probably around 3500 BC in Mesopotamia; see chapter 2). Scribes, being both learned and skilled, occupied a place of power and influence: Only they could encode language in seemingly unintelligible marks scratched on vellum or onto clay tablets. Only they could transmit language—orders, myth and legend, commentary, narratives, accounts—by means other than voice and drawing. They occupied this position of unique authority, essentially acting as a powerful intellectual elite, for some three thousand years.

It's difficult for the modern mind to appreciate the extent of the power wielded by the scribes of that time—and by their royal masters. They completely controlled communications. Among other things, this meant that they could effectively suppress (or incite) rebellion; after all, one cannot successfully revolt if one cannot communicate one's revolution. As William A. Katz notes in *Dahl's History of the Book*, "[F]rom time to time groups revolted, but lacking any real way to pass messages other than by voice, the rebels were quickly subdued. A modern proximity to the situation is found, of course, in totalitarian states."

Unsurprisingly, if you control media, you control the people. And the scribes controlled the media of the time.

Papyrus

Some of the first items that we might recognize as books were the papyrus scrolls used in the Near East. The stem of the papyrus plant was used to make sheets of paper-like material that was formed into rolls of from twenty to thirty feet in length. Text was written on the surface, which was about twenty inches wide. The scroll was stored on a spindle, and the reader unrolled the "book" as it was read, taking it up on a second spindle. (We have few extant examples of early papyrus scrolls, given the fact that the material was easily damaged. Dampness was especially ruinous for papyrus, though a few—some of them stored in graves in the parched deserts of Egypt—have survived.)

It was mostly books in this form that were stored in the celebrated library at Alexandria. Founded in 332 BC by Alexander the Great, the city itself was known as "the Paris of the Mediterranean." The "library" was actually quite a bit more than that: It was a think tank of sorts, and a university, an educational foundation, and more. Euclid and Aristophanes, among others, studied and taught there.

The goal of the librarians at Alexandria was simple and startlingly ambitious: They wished to collect in one place the then-current texts. *All* of them. In all languages. And from all cultures. Even in those days, that would have meant a *lot* of books. In fact, the laws of the port city proclaimed that any arriving ship could be searched, not for contraband or taxable goods or spoils, but for books; any books discovered that were not already in the library were confiscated. The library may have housed as many as 700,000 scrolls. In today's terms, says William A. Katz, that means that there were about 70,000 to 80,000 specific titles in the library—and multiple copies of many of those.

In 48 BC, the library at Alexandria was burned, and many of the books and possibly much of the library itself was destroyed. But the library's real end may have come in AD 391 at the hands of Christian zealots acting in response to decrees that had outlawed "pagan practices." (And a pagan book was thought to be . . . well, a pagan practice of sorts.) Alternatively, and somewhat ironically, it may instead have been destroyed by *Muslim* zealots about one hundred years later. So, the library may have met its end at the hands of one zealot or another. (The story goes that a victorious Muslim general was asked for the remaining books. When he wrote to the caliph for instructions, the caliph, Omar, is said to have replied, "If those books are in agreement with the Quran, we have no need of them; and if these are opposed to the Quran, destroy them." Either way, the books were apparently deemed incidental, insignificant, and unnecessary.)

In the end, we simply don't know for sure what happened to the library at Alexandria. All we know is that over time it was lost, and with it were lost documents of inestimable cultural and historical value.

The next major step in the evolution of the book came from the Romans. Introduced as early as the first century AD, the codex—a stack of sheets glued on one end and bound using a cover somewhat thicker than the paper

This image of the burning of the library at Alexandria is from an early-twentieth-century history textbook. *Image in the public domain.*

itself—had by the sixth century AD become the dominant format used for collecting and presenting information in books. With some modifications, this is in fact the form of the book as we know it today, including what may be the most famous book in the world.

The Gutenberg Bible

As everyone knows, Gutenberg's Bible heralded a fundamental transformation in printing, and indeed, in communication. That transformation happened slowly at first, but picked up speed very quickly. (Note that the Bible may not have been the first thing printed by Gutenberg—there is evidence that a book of psalms and a series of receipts for Church indulgences may have predated his Bible—but that book is most likely the first complete, major work to come from his press.)

We know that Gutenberg's Bible was in press—though not yet bound and perhaps not even gathered in quires (sections)—by March of 1455 at the latest. There were around 180 (some sources say 120) copies of that first Bible (the forty-two-line Bible, so-called because forty-two lines of text in each column were printed on a page) printed over a period of about five years. That may not sound like much today, when we can use huge computer-controlled presses to print thousands of copies of a book in an hour, but for the mid-fifteenth century, it was nothing short of amazing.

Prior to the invention of movable type, it might take a skilled scribe years to create one illustrated and illuminated copy of a Bible in codex form; certainly, a single scribe—no matter how skilled—could produce no more than one or two books over a period of a year, even if he managed to devote almost every waking hour to the task. (Of course, the early scribes were often monks and had other duties. Many times, they had to make their own quills, inks, and pigments. Then there were prayer, meditation, gardening, and other chores; attending to these other tasks extended the length of time required to create a Bible—or any book or pamphlet, for that matter.)

About forty-six copies of the Gutenberg Bible still exist, and you can see them at various museums, libraries, and universities. (In the UK, for instance, copies can be viewed at the Lambeth Palace Library in London, and also at Cambridge and Oxford. In the United States, copies can be seen at the Morgan Library & Museum in New York, the New York Public Library, and at the University of Texas at Austin.) They are incredibly valuable; the most

This copy of the Gutenberg Bible, printed on vellum, is pictured at the New York Public Library in 2009. The red introductory lines were added by hand after printing, a process called *rubrication. Image licensed under the Creative Commons Attribution-ShareAlike 2.0 Generic License.*

recently auctioned copy of a complete Gutenberg Bible fetched over $2 million in 1978, and they are said to be worth much more than that now. (A Gutenberg Old Testament—that is, an *incomplete* Gutenberg Bible—sold about ten years later for well over $5 million.)

Another Start-up Fails

Yet, if you were to visit one of the universities or libraries with a Gutenberg Bible in its collection, and if you were to examine one of those remaining copies, you might notice something odd about it. Something is missing. Even if you were allowed to pick up and leaf through a Gutenberg Bible (and there is no way to communicate just how unlikely a scenario that is), you would not find anywhere in or on that Bible something you would surely have expected to discover: Gutenberg's name.

The absence of Gutenberg's name is not an error or an oversight; it is instead the result of a lawsuit. By the time the Bibles were printed and bound, none of the books belonged to Gutenberg. Nor did the press on which they were printed, nor the building that housed the press. In fact, by the time the Bibles were ready for sale, all of the supplies and equipment associated with the venture—including the actual printed product—belonged to Gutenberg's main investor, a man named Johann Fust (sometimes rendered as Faust).

Fust, who had between 1449 and 1453 loaned Gutenberg the then-considerable total of 1,600 florins (about 800 of today's dollars), had become—as sometimes happens with investors in start-ups—disenchanted with the delays and with the inventor's seeming inability to deliver the product he had promised. Fust eventually sued Gutenberg. Gutenberg, a brilliant innovator but perhaps not the best businessman, lost the suit and was required to pay all interest due and to return any unspent capital. Fust now owned everything: Gutenberg's materials, supplies, drawings, test prints, type, and the press itself. Fust immediately partnered with Peter Schoeffer, formerly Gutenberg's foreman (that had to hurt) and a Fust protégé, creating a printing firm that would stand for many years. (Schoeffer eventually married Fust's daughter, so there may have been much more to the story.)

In 1456, Fust and Schoeffer—not Gutenberg—published the famous forty-two-line Bible. So the world-famous "Gutenberg Bible," though certainly produced by Gutenberg on a press of Gutenberg's design and with type of Gutenberg's own making, was arguably a product of the firm of Fust and Schoeffer. In any case, in spite of the fact that we call it "Gutenberg's press," and notwithstanding the many references (then and now) to his role in the invention of this form of mechanized printing, Gutenberg's name does not appear on or in any of his famous Bibles.

Let's Just Keep This to Ourselves, Shall We?

Just as the scribes of ancient Mesopotamia constituted an intellectual elite, later scribes creating manuscript Bibles in Latin were also members of a privileged and influential group. Both groups were used by the reigning political and ecclesiastical powers (and recall that quite often these were one and the same) to keep a tight rein on the populace.

Like the scribes, the priesthood exercised power in the name of the Church, and together with whatever secular authorities were in office, they tended to control the "media" of the day: The Bibles and other books created by scribes (and later by pressmen) were produced and distributed mainly under license from those authorities. And the Church, in particular, was not in favor of widespread literacy on the part of the laity.

This was especially true when it came to Bibles, because the Church feared that allowing just anyone to read the Bible could lead to individuals interpreting (and misinterpreting) the word of God however they pleased. Thus, "efforts were made to restrict the reading of the Bible by the laity, since its use seemed the source of medieval heresies," as Williston Walker has noted. Some sources have said that in 1229, for instance, a Church council in Toulouse, France, explicitly forbade anyone who was not a priest from owning a Bible.

Of course, local councils did not generally speak for the Church at large, and any "local" banning may have merely been a somewhat heavy-handed response to local heresies, bad translations, etc. But it does seem that the theocracy as a whole did often censor or ban versions of the Bible—as well as other writings—that had not been approved by the Church.

In 1502, the Holy See ordered the burning of all books that could be interpreted as questioning papal authority. In 1516, the Fifth Lateran Council approved a decree that forbade the printing of *any* book without the Vatican's approval. And the *Index Librorum Prohibitorum* of 1559 makes it clear, says Tom Heneghan, writing in *America: The National Catholic Review*, that the Church fathers were uncomfortable with the notion of translating the Bible into "the common tongues," a prohibition which, given that the common people did not speak Latin, might have had an effect similar to an outright ban. (Not surprisingly, the *Index*, Heneghan points out, also banned all books by Luther, Calvin, and other Protestant reformers. He noted, "Since translating the Bible into vernacular tongues was a Protestant specialty, all Bibles but the Latin Vulgate were banned. The Talmud and the Koran were also taboo.")

Keep in mind, though, that the medieval Church was not simply exercising an arrogant, despotic desire to restrict knowledge merely because the Church fathers feared subversion. Remember that the Church was a sociopolitical body intent on creating an integrated world, a unified and all-inclusive realm in which everyone worshipped the same God in the same way. (The word *catholic*, after all, means "all-embracing.") The Church therefore saw itself as a force for good, and millions saw it in the same light. Thus, it sought a unified Christendom with a single language—Latin—so that the world could experience the peace that it believed it offered.

Restricting the Technology

Whatever the reasons, the technologies—and the fruits of those technologies—of early "mass media" were largely withheld from most people in at least three ways.

First, as we have seen, what books or other printed materials did exist had to be approved by the authorities, and their publication could be summarily forbidden by those authorities if the materials were deemed potentially subversive. (And as we know from our experiences with totalitarian regimes throughout the ages, any material that did not explicitly praise the regime could be—and most likely would be—deemed subversive. This is true regardless of whether the regime in question is ecclesiastical or secular in nature.)

Literacy was also an issue. Even if someone were exposed to seditious literature (or any other type of literature), it couldn't have much of an impact if he were unable to read. And in those days, unless one were trained to become a professional (a lawyer, priest, engineer, etc.), it is unlikely that education was high on anyone's list; bringing in a crop and hauling stone and making brick or cloth were certainly more important to most farm workers and laborers than learning to read and write. (Even now, the overall literacy rate is startlingly low in many developing countries: Pakistan's is 55 percent, Kenya's is 72 percent, and Afghanistan's is 32 percent.)

A commoner during the fifteenth and sixteenth centuries almost certainly would not learn Latin, which was the only language in which the Bible and many other materials were allowed to be printed. In the end, the growth of printing was tied inevitably to increases in literacy; which one caused the other is a matter of some debate. (Some scholars have argued that printing did not kick-start a widespread shift toward literacy, and that literacy was in fact already on the rise. As Brown University professor Carl F. Kaestle points out, "By the time of the invention of moveable typographic printing, literacy levels were relatively high, literacy was spreading, and books were known and widely distributed. The list of books in circulation before the mid-fifteenth century is long and includes the main writers of classical antiquity.")

It's also worth noting that, largely because of the Church's attitude toward literacy and the dissemination of books—and also because people clung tenaciously to their earlier beliefs—the medieval populace remained mostly ignorant of the Bible in any case, and almost completely unaware of the underpinnings of the religion in which they professed to believe. As William Manchester comments in *A World Lit Only by Fire*:

> *Although they called themselves Christians, medieval Europeans were ignorant of the Gospels. The Bible existed only in a language they could not read. The mumbled incantations at mass were meaningless to them. They believed in sorcery, witchcraft, hobgoblins, werewolves, amulets, and black magic, and were thus indistinguishable from pagans.*

So, the Church fathers certainly had their work cut out for themselves.

Finally, there is the matter of cost. At first, books were the province of the wealthy. A manuscript book might sell for several times the annual pay of a laborer. Books were valuable enough to have been chained to desks in universities and libraries, and only the wealthy could afford a single book, let alone a library of them. Even a printed book of the sort created on Gutenberg's (or a similar) press cost enough that one had to be relatively well off in order to consider buying one, at least in the early days of the printing press.

But that didn't remain true for long.

More-Affordable Printing

Eventually, as we have seen, new technology tends to trickle down to the less powerful until it ultimately becomes ubiquitous and affordable. Almost always, the net effect is a leveling of the playing field: Books are now relatively cheap, and the information (or entertainment) in them is available to just about everyone. Knowledge truly *is* power, and that knowledge—and, thus, that power—is now much more readily available to those who seek it.

Gutenberg's press helped to democratize printing, and any constraints imposed by the government and by the Church did little to put a halt to the publication and dissemination of religious and secular works, especially once the printing press and movable type had been invented. Technology, as we have seen, can only rarely be withheld from the public, and then not for long. And attempting to stop the flow of information in particular is a very difficult—perhaps ultimately impossible—task.

By 1640, there were by one estimate some twenty thousand different titles (titles, not copies) available in Britain alone. Books were becoming more and more affordable, and commoners were more and more often able to read news and commentary in their own tongue, even if they did not understand Latin.

As Simon Horobin explains in Alexandra Gillespie's *The Production of Books in England 1350–1500*, the reason that books were expensive before the advent of printing is obvious. Equally obvious is the reason the prices dropped so precipitously after printing came along:

> *The production of manuscripts was [at first] largely a bespoke trade: a client would approach a scribe or stationer and order a copy of a work which would then be produced for that single client. Printing brought with it speculative production, whereby large numbers of copies of a work were printed and then circulated to a much wider market. The cost of such books also decreased, allowing books to be owned by a wider cross section of society.*

The Effects of Printing

Books—and their ready availability—can be said to have laid the foundation for many of the intellectual, political, and religious changes that would take place during the following centuries. There is thus an argument that there has never been a more significant, more impactful invention than that of printing via movable type.

The availability of printed books and pamphlets may even be said to have aided the Reformation by making it possible to produce and disseminate the pamphlets, books, and ballads that comprised the propaganda that supported the Protestant cause.

After all, the Reformation was spearheaded by the printed (and spoken) words of a former Catholic monk who had determined that the Church, magnificent though it was, had lost its way. When Martin Luther nailed his "95 Theses" to the Castle Church in Wittenberg (in those days, a standard way of announcing an upcoming sermon or debate), he turned on his own Church, criticizing, among other things, practices that had made the Church wealthy, including the sale of indulgences.

These are the doors at the church in Wittenberg upon which Martin Luther is said to have nailed his "95 Theses." *Image licensed under the Creative Commons Attribution-ShareAlike 3.0 Germany License.*

But Luther was not simply a firebrand and a revolutionary; he was a firebrand and a revolutionary who had happened along at exactly the right time to take advantage of the latest technology available to help him spread the word—*his* word: the printing press. It's no accident that Luther's tracts were printed in Germany (and in German, not Latin, so as to ensure the widest distribution), and on presses not very different from those created only sixty or so years earlier by Johann Gutenberg. Within a few months, all of Europe was awash with copies of the "95 Theses."

In response, and somewhat unwisely, it seems in retrospect, Pope Leo X excommunicated Luther and condemned his works, ordering that they be burned. Naturally, the immediate effect of the Pope's Bull of Excommunication was to increase the sale of pretty much everything that Luther had ever written. Even the demand from other countries skyrocketed. It may have been an early example of what has recently come to be called "the Streisand Effect," referring to the unintended consequence of further publicizing information by trying to have it censored. (In 2003, singer Barbra Streisand tried to suppress photos of her Malibu estate, which merely served to publicize the existence of the photos; thus, instead of an occasional lonely web wanderer accidentally happening upon the collection—intended to document coastal erosion, not a particularly rousing topic, one could safely assume—her lawsuit generated so much publicity that the collection was for a time seen by hundreds of thousands of visitors per month.)

Barbra Streisand, still active in the arts and in liberal causes, is seen here at the Clinton "Health Matters" Conference in La Quinta, California, in 2013. *Image licensed under the Creative Commons Attribution-ShareAlike 2.0 Generic License.*

Thus, the Reformation caught fire partly because that fire was fueled by books and pamphlets printed using the technology created by a former goldsmith of Mainz. (And so many were printed that even today, some 630 of these tracts remain, even though they were printed almost five hundred years ago.)

Relatively quickly, advances in printing technology had an incredible effect on literature, on art, on religion, on thought.

The University of Nebraska's Stephen Buhler notes that the transformation was both sudden and stunning, observing that within a couple hundred years we became a society—and a literary culture—increasingly dominated by print. "That's scarcely as swift as the digital revolution, but it certainly paved the way for future technologically-driven transformations," he says.

The Church was right to worry that the ready distribution of books might result in the spread of what it viewed as sedition. As Lucien Febvre notes in *The Coming of the Book: The Impact of Printing 1450–1800*:

> *[I]f it does not succeed in convincing, the printed book is at least tangible evidence of convictions held because it embodies and symbolizes them; it furnishes argument to those who are already converts, lets them develop and refine their faith, offers them points which will help them to triumph in debate, and encourages the hesitant. For all these reasons, books played a critical part in the development of Protestantism in the sixteenth century.*

Books are, among many other things, invitations to revolution. It would be quite a stretch to say that the printing press *caused* the Reformation, but it is certainly true that it helped the movement along. One author called the Protestant movement "the first mass-media event in history." That movement may also have been the first media-mediated revolution, a distant but direct ancestor of a seismic shift in public opinion that would occur hundreds of years later, when a war raging in the jungles of southeast Asia was broadcast almost in its entirety into the living rooms of ordinary people who might otherwise have had little knowledge of it, and thus, no opportunity to protest its prosecution.

New Printing Technologies

As we would expect, the printing industry has not been unaffected by technological advances. In fact, printing has been subject to an ongoing process of revolution ever since Gutenberg, and that revolution continues today.

For centuries, the basic mechanics of the letterpress remained almost unchanged since Gutenberg's day, but in the early nineteenth century, travel and industry in general were changed forever with the advent of steam power. And, not surprisingly, the steam revolution affected the printing industry just as it affected every other facet of manufacturing. In 1813 John Walter, a former coal merchant who had become the publisher of the *London Times*, purchased and installed two double presses worked by steam. The new presses were so much more efficient that he could afford to print enough copies of the paper (and could print them quickly enough) to service what was for the time a huge audience. (He also made so much money that he was able to refuse bribes and ignore political pressures, which made the *Times* different than most other papers of the day.)

The new presses were quite expensive, though; James Perry, publisher of a rival London paper, refused to consider spending that much money, believing that the presses cost more than any newspaper was worth. (This is a mistake we will see time and time again when it comes to judging the affordability or efficacy of a technology. The fact that it is currently quite expensive does not mean that it will remain so. In fact, it most assuredly will *not* remain expensive; technology tends to become affordable—that's one of the ways in which it trickles down and begins to level the playing field for the masses.)

The Linotype and Offset Printing

But steam power was just the beginning; like all industries, the printing industry was to be transformed time and again by technological developments.

One major advance came in the late 1800s, with the invention of the Linotype machine. An incredibly complex electromechanical contrivance (Thomas Edison called it "the eighth wonder of the world"), the Linotype took Gutenberg's revolution and revolutionized it all over again, allowing operators at a keyboard to type sentences that were then formed into whole lines of type cast in lead, each line molded on the fly within the machine itself. The machine used a precise choreography of wheels, belts, levers, slides, and gears to create lines of type (called "slugs") that were automat-

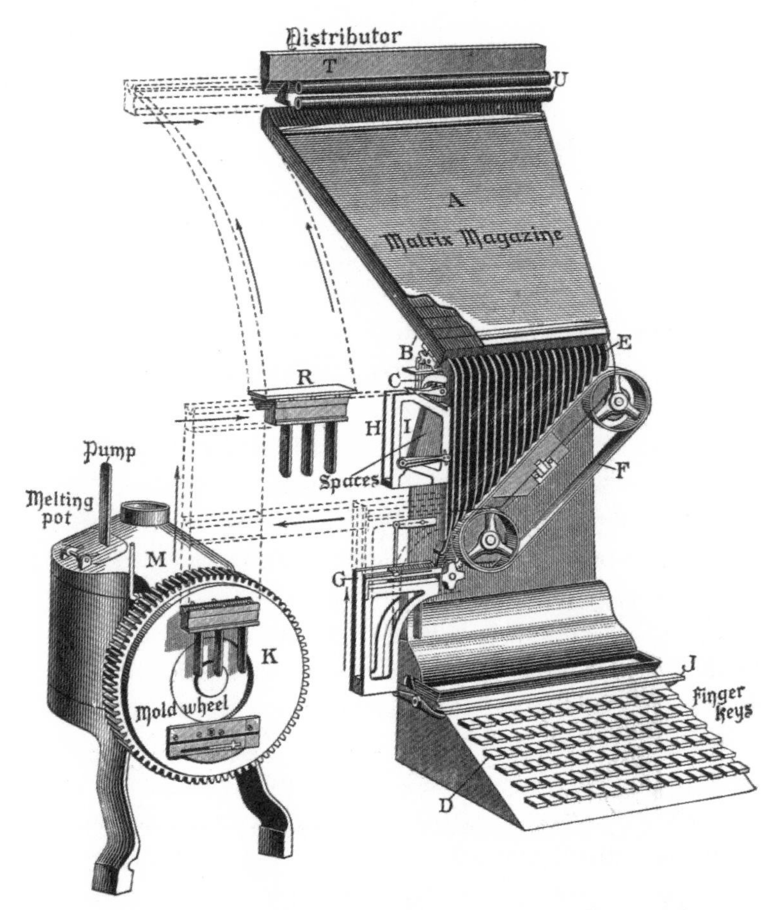

This Linotype machine, circa 1904, is typical of the era. The complex system served its purpose well for over half a century, enormously speeding up the process of typesetting. Note the heated pot in which lead was melted and then molded into "slugs" of type. *Image in the public domain.*

ically moved into position in order to compose the page for printing, and which could be melted down later for reuse. (Skillful Linotype operators were prized employees, partly because correcting mistakes was difficult and time-consuming; essentially, the entire flawed line of type had to be discarded, retyped, and recast. It was better—much better—to avoid making mistakes in the first place, so accomplished Linotype operators who made few mistakes were very valuable.)

Operating a Linotype took real skill, but it did not require the sort of artistic gifts required to create reversed-letter matrices, and composition was much faster on a Linotype than even a skilled compositor could manage when placing one letter at a time in a chase. The Linotype machine was so successful that it was still being used in print shops and small newspapers as recently as the 1960s and 1970s.

But then came offset lithography. Actually an older technology into which was breathed new life by advances in chemistry and photography, *lithography* literally means "printed from stone," and originally, that's exactly what it was: A design was sketched on a porous stone (limestone was popular), and then the stone's surface was wetted. The design area (usually an image of some sort) repelled the water. Ink was rolled onto the stone, but did not spread on the moist (blank) areas. When paper was pressed against the stone, it left a clean impression of the image on the paper.

Lithography as a process became quite popular, especially once steam-powered litho presses became available in the 1850s. At first, these also used stone for the source images, which were transferred to a blanket-covered cylinder; it was the cylinder which then made contact with the paper—thus the term "offset," used to describe a process in which the image was transferred to some other medium before being transferred to the paper.

Nowadays, we use photochemical processes to create images (including images of text, of course) on a plate, and the plate (which acts like the stone in the original lithographic process) is inked and transfers the image to a rubber mat; from there the rubber mat places the image onto the paper.

Offset lithography is still the most common form of printing, but there were more revolutions in store for the printing industry.

The Digital Age Arrives

In a brief 1976 editorial on the subject of computerized typesetting (who knew that this was a subject about which one might feel compelled to write an editorial, brief or otherwise?), C. A. Lang marveled at advances made in the computerization of publishing. In his conclusion, he noted that, with the advent of home and small-business printers, writers could now deliver camera-ready copy to a publisher. "How long," he wondered, "before we are able to accept an article in machine-readable form?"

The answer was, of course, "not very long at all."

Ever since World War II, governments had been investing in building more and more powerful computers, and in finding ways to apply that computational power to various tasks, military and otherwise. One of those avenues of research turned out to be typesetting. After all, governments collect, store, analyze, and distribute enormous amounts of data. Getting that data into computerized storage and then finding ways to print it when and where it was needed was a task of no small importance.

Nonmilitary public agencies followed suit, of course, and as costs dropped and new technologies were developed, private industry (which had in any case always functioned as government-contracted partners working in these areas) joined in. It was obvious that, sooner or later (and most likely sooner), computers would find their way into the printing process. (But not obvious to all. One university professor in the early 1970s determined that using computers to compose and print would always be too expensive, and he had reams of calculations to prove it. And the consultant who published those calculations felt that the professor himself was actually being "unduly optimistic.")

Digital Typesetting

Early photocomposition systems utilized analog electromechanical technology, but soon began to include punched-tape inputs used to position a rotating disc such that a flash of light exposed photographic film, thus generating an image of a character. By the 1970s, though, photocomposition had become a wholly digital undertaking, and information about character, character size, placement, and spacing could be transmitted via keyboard, causing a cathode ray tube to emit pulses of light, exposing film much as the rotating disc had done somewhat earlier.

Then the digital skies let loose, the dam broke, and a full-blown digital printing revolution was upon us. According to the University of Florida's E. Haven Hawley, improvements in digital memory and storage paved the way for that revolution:

> *The ability to store information in locations other than the printing facility moved tasks like typing in data "upstream" to editors and designers following the diffusion of computer technologies into those occupations in the mid-1970s and 1980s. Digital technology in those decades allowed*

The Linotype CRTronic 360 was a 1970s- and 1980s-era "cold type" photo compositor descended from the original "hot lead" linotype machine. It stored data on single-sided 5.25-inch floppy diskettes. *Image licensed under the Creative Commons Attribution-ShareAlike 2.0 Generic License.*

> *information to be transmitted by disk or modem, without rekeying text or transporting volumes of punched paper tape. The availability of personal computers, the prevalence of computers in many work environments, and the development of software that [integrated] text and graphics into a single file reshaped production technologies. Increased memory, software that allowed users to position text and graphics together on a single page (called pagination), and the availability of computers ranging from sophisticated pagination systems to simpler personal computers in homes transformed typesetting in that decade, removing many of the diffusion barriers inherent in analog technologies.*

Along with everything else, printing and publishing have been greatly affected (some might say *redefined*) by the computer age. By the late 1970s and early 1980s, reporters at even small daily newspapers were entering stories into computerized systems. Soon after, computers would take over pagination, allowing editors to lay out the entire issue onscreen before committing it to what continued to be called "typeset," in this case actually sending the pages digitally directly to the press itself. Gone were the days of paste-up tables, "hot lead" (a term derived from the lines of lead type cast in Linotype machines), and frenzied, fedora-wearing reporters shouting for copy boys to deliver sheets of yellow foolscap to editors and layout men. (Gone too were the rooms filled with the clackety, ratcheting sounds of typewriters being pounded upon by deadline-stressed writers hurrying to finish a story before press time. Now such rooms tend to be eerily quiet; an occasional telephone might ring, chime, or otherwise announce a caller, but today's reporters—just as stressed as always, of course—sit at whisper-quiet keyboards, peering fixedly at LCD monitors while the clock ticks and beads of blood form on their foreheads. Actually, clocks rarely "tick" these days, but you get the idea.)

After computerized typesetting came to large commercial presses, it came to our home and small-office workspaces via what's called *desktop publishing*, or DTP. By the late 1980s, it was possible for home users or small businesses to use DTP software to create anything from a flyer to a full-blown book. With the advent of laser printers, one could in fact deliver camera-ready copy for most types of publications, and one could certainly, in answer to Lang's earlier question, deliver a publication to the publisher "in machine-readable form."

(The first DTP system that I recall using was called Ventura Publisher. It appeared on my desk one day in the form of a bulky package that contained thirteen—that's right, *thirteen*—3.5-inch diskettes and an intimidatingly large manual. It was incredibly sophisticated for its day, but nowhere near as powerful as today's QuarkXPress or InDesign systems. Still, there was no denying that a revolution in publishing had arrived. It was with an immense sense of power—and also feeling the weight of a certain amount of responsibility—that I realized I had sitting on my desk a tool that reached back to Alexandria, Gutenberg, and Luther, and which allowed me to create, to publish. Of course, the fact that I was writing users' manuals for educational software aimed at high school students tended to deflate my ego a bit; I was not, apparently, going to set the world afire and precipitate another Reformation. Which was probably just as well, really.)

Digital Books

But we're not quite finished talking about new technologies and their effects on publishing and printing. (In fact, we probably never *will* be finished.)

When I travel, I read a lot. On the plane, in the car, relaxing on a relative's back porch or a hotel balcony, even when dragged along on a shopping trip to visit those quaint (read: expensive) little shops—almost everywhere I go, I have a book with me. It might be work-related, it might be for research, it might be something I'm reading for pleasure. In fact, when I pack my bags for a trip, I generally include well over one hundred books tucked in among my jeans, shirts, and toiletry kit.

But of course I don't actually carry one hundred printed books; together, those could easily weigh well over seventy-five pounds, and take up my entire suitcase. Instead, I carry the digital versions of those books, e-books, which weigh essentially nothing. There are several excellent dedicated readers available (the Amazon Kindle and Barnes & Noble Nook readers stand out, of course), but I like to use a small general-purpose tablet, since I already have it with me for e-mail, music, web surfing, and note-taking. (Right now my go-to tablet is an iPad mini. It's about the size of a paperback book, though much thinner, and weighs about three-quarters of a pound. I can easily store hundreds of books on it.)

Not only do I save weight, but I also save a good deal of money when I buy the digital version of a book. The large-format paperback edition of one book I recently read (Erik Larson's enthralling *Dead Wake*; Larson is an incredible researcher and a gifted storyteller), costs $17 from Barnes & Noble. The e-book version of the same publication costs $11, or about 34 percent less than the printed book. (If you purchase one hundred e-books and save two or three dollars apiece on them, you will have just paid for your tablet or reader. Of course, if you borrow digital books from a library, your savings are even more impressive.)

I also appreciate the authorial advantages offered by digital text. I can search for a term, a name, or a phrase; I can highlight and make notes, and I can then list all of those highlights or notes together in one place. If I'm quoting a passage, I can copy and paste that passage into a manuscript, thus ensuring that accessing the quote is convenient and that the quote itself is accurate. Although it takes some getting used to, there are undeniable advantages to using digital books.

Dr. Stephen Buhler is the Aaron Douglas Professor of English at the University of Nebraska at Lincoln and the author of *Shakespeare in the Cinema: Ocular Proof* (Albany: State University of New York Press, 2002). Dr. Buhler often discusses in his classes technology and its effects on literature and communication. *Image courtesy of Steve Buhler. Photo by Michelle Zink.*

Now, having said all of that, I should point out that I do realize not everyone is comfortable with e-books. Many of us (including myself; I am an English teacher, after all) enjoy the experience of handling and reading from a printed book; we feel that although the informational content may be the same, the *experience* of reading an e-book does not quite measure up to that of reading the printed version. The fact of the matter is that we *love* books, and printed books are what we grew up loving.

And there is a great deal to be said for the love of books—physical, printed, bound, ink-on-paper books. The inexplicable but undeniable beauty of them, stacked in disordered piles, or arrayed neatly on a shelf, or scattered about the house on every horizontal surface. The heft of them. The musty, paper-y smell of old books and the sharp, fresh-ink-on-paper smell of new books. The subdued colors of the bindings and the flash of cover photos. The tactile and aural experience of flipping a page. The convenience of scribbling in a margin or using a brightly colored sticky note to mark a favorite spot. (And some of you, I daresay, dog-ear pages in books. It's okay, you can admit it. I've done it. But when I do, I can hear the angry voice of my mother: "Books are important! You don't treat them that way." And Mom was right, of course.)

I know (and agree with) all of that, but it's difficult to ignore the ecological and economic imperatives that are driving the adoption of e-book technologies. Information, after all, is a weightless, formless commodity. For centuries, the best way to share that information was to attach it to a great deal of weight (in the form of paper), and then pay to ship that weight all over the world.

That is no longer the best way (that is, the most efficient, least harmful way) to communicate, to share information; it's no longer the best way to show people how to do things, or to explain the world to them. Nor is it the most efficient way to allow people to share in the breathtaking adventure that is *Moby-Dick*, or to enjoy the whimsy of *Peter Pan* or the biting wit of Shakespeare; it's no longer the least-expensive, most-accessible way to get caught up in the excitement of the latest political thriller, the currently popular young adult fantasy, or the most recent medical mystery.

Printed books, as much as many of us love them, are becoming less necessary because we now have alternatives to them that are more affordable and less injurious to the environment. Will physical books go away? Someday, perhaps, though surely not for quite a while. Still, I can see (and not without some profound misgivings) a new Middle Ages for books: a future in which printed books are once again so rare and so expensive that they are the province only of the wealthy, displayed for their beauty (and, of course, to advertise their owners' affluence), but chained in place because their loss would be financially catastrophic.

The University of Nebraska's Stephen Buhler doesn't quite agree, though. Dr. Buhler feels that the printed book is here to stay, in some form or another, even as digital books increasingly make sense when presenting and discussing certain types of material.

"I foresee ways of managing 'print-on-demand' that keeps costs low without economies of scale. I also foresee different kinds of books suited to different technologies. The e-book is ideal for . . . [multimedia] presentation. A lot still has to be resolved over issues of Fair Use and intellectual property, but I would love to see books of criticism devoted to, for example, film or music that provide samples from (or links to) all the works under discussion. As much as I love the traditional book and as firmly as I believe in its continuance (and some studies suggest its revival is already under way), there are some things that e-book technology can do so very much better." (If Dr. Buhler is right about print-on-demand technologies, perhaps printed books might once again become what Simon Horobin earlier called "a bespoke trade," only this time a much more affordable, more accessible one.)

If the printed book does become a rarity, it will be a sad time, and we will have lost something important—something magnificent, in fact. But we will *not* have lost—in truth, we will have greatly enhanced—the ability to transmit information, to communicate with readers; the power that derives from knowledge will be available to *more* people, rather than to fewer.

Failure: The Best Route to Success

Sometimes technologies change lives. Sometimes, they change the world.

In the context of this book, Hawley's earlier use of the term *barriers* is telling. This chapter—in fact, this entire book—is about technological changes that remove barriers. Publishing was at one time an undertaking abounding in barriers—legal, technological, religious. At first it was impossible to publish because it was forbidden. Later, it was merely horrendously (and dishearteningly) expensive. Until recently, serious barriers to entry stood in the way of those who would publish. It cost millions of dollars to purchase and house and maintain a press and to hire people to operate it. It cost a good deal of money and took much time and skill to create, design, and typeset a publication.

For better or worse, that is no longer true. Many of those barriers have disappeared, or at least been mitigated; they no longer pose serious obstacles to those who wish to write, to print, to publish. It no longer takes a huge investment, nor does it take the skills of a lithographer, a Linotype operator, or a pressman to publish a tract, a magazine, or an entire book. (As always, this is both good and bad, depending on one's perspective. Some may argue that such barriers served a useful purpose, helping to ensure that access to the process was limited to those who could convince others of the worth of what they intended to publish. But there's not much to be done about that; when we democratize technology, we democratize it for *everyone*, even those whom we would rather have denied access to that technology.)

As will be noted many times in this book, no one—not even its inventors—knows where a technology will take us. People will use technology in unintended ways; the tools will themselves turn out to have unintended consequences; and as often as not the new invention will be used to create still more tools and technologies, the likes of which could never have been anticipated by the initial technology's originators. Thus, the terrific (and frequently noted) irony that Gutenberg, by all accounts a devout Catholic, should have invented a device that would be used so effectively to further Martin Luther's Protestant Reformation. Luther himself said that the printing press was "God's last and greatest gift." It certainly helped the fiery preacher promulgate his sermons, his tracts, and his attacks on the sale of indulgences (which he called the Church's "[d]isgraceful Profanation of the Lord's Supper, by making a common sale of it"), galvanizing the growing protest movement that eventually became Protestantism itself. One assumes, of course, that Gutenberg would have been horrified at the use of this "gift" by those who would profane, protest, and ultimately abandon the Church he loved.

Gutenberg's enterprise may have been among the earliest of start-ups, but even back in the fifteenth century it conformed to our contemporary perceptions of how a start-up typically operates: The man spent money he did not have in order to create technology he couldn't guarantee would work so that he could sell a product he wasn't positive that anyone actually wanted.

And he *failed*, at least from the narrow perspective of what constitutes a successful business: His company was taken from him; he made very little money. His name appears on no books of the era, not even on the ones printed on his own press. Johann Gutenberg created a device—in fact, an industry—that transformed (and continues to transform) the world; there's a pretty powerful argument that Gutenberg's printing press may be among the most important technologies ever developed, in which case Gutenberg may have been the most successful failure in history.

And we're about to meet another successful failure. Almost 450 years after Gutenberg, Henry Ford would eventually succeed in spectacular fashion, but his first two business ventures failed miserably. Beginning his career in 1891 as an engineer with Thomas Edison's Edison Illuminating Company, he didn't start Ford Motor Company until he was forty years old, and didn't revolutionize the assembly line–driven manufacturing process until ten years after that. Ford's spectacular failure eventually became a stunning triumph, but his success was by no means guaranteed.

Chapter Four

Goin' Mobile: The Automobile

If we had a reliable way to label our toys good and bad, it would be easy to regulate technology wisely. But we can rarely see far enough ahead to know which road leads to damnation. Whoever concerns himself with big technology, either to push it forward or to stop it, is gambling in human lives.

—Freeman Dyson

The Short, Unhappy Life of Henry Bliss

There have been few men more egregiously misnamed than Henry Hale Bliss.

To begin with, Henry's promising future as a husband and father was painfully derailed when his young wife died only three years into their marriage, and two years after having given birth to their son, George.

Bliss married again, this time to the former Evelina Matilda Davis Livingston (of New York's wealthy, high-society Livingstons), but for reasons that remain murky, that marriage ended in divorce. (Still, Bliss and his former wife remained close and visited one another fairly frequently.)

Then, in August of 1895, fate dealt Bliss what one would have surely thought was its cruelest blow. Bliss's stepdaughter, Mary Alice Livingston Fleming, was arrested and charged with murder, having apparently sent *her* ten-year-old daughter to deliver poisoned clam chowder to Fleming's mother, Henry Bliss's former wife. Investigators believed the murder was a devious and coldhearted (and ultimately successful) attempt to secure access to Evelina Bliss's fortune, which Mary Alice stood to inherit on Evelina's death. The chowder was laced with both antimony and arsenic—as if Mary Alice couldn't quite decide which would do a better job and had determined to employ both, just to be sure. *Any job worth doing*, as they say.

In the end, according to a chemist hired by the coroner, it was the arsenic that did in poor Evelina—and not surprisingly, as the soup apparently contained enough of it to kill her several times over. (Had Evelina's young daughter chanced to sneak a little of the chowder during her errand, the child would surely have died.)

To add even more drama to the already sordid goings-on, Mrs. Fleming was actually arrested *while attending her mother's Manhattan funeral.* (Mary Alice was never actually married to Mr. Henry Fleming, a well-to-do oil merchant, but won the right to use his surname during an especially nasty 1883 breach-of-promise suit, in the course of which Henry accused Mary Alice of prostitution. In June of 1896, Mary Alice would be acquitted of the murder, possibly because—and here's a neat piece of irony—her inheritance from the dearly departed Evelina Bliss provided her the wherewithal to hire Charles W. Brooke, one of the country's top criminal lawyers, to represent her.)

The story was a gold mine for New York's sensation-seeking newspapers, combining as it did the most scandalously lurid aspects of high-society life: murder, sex, and money. What more could the newspapers of the time—or of *any* time—have wished for?

Bliss, a real estate broker in New York City, had now lost his first wife, his former wife, *and* his stepdaughter.

But fate was not yet finished with the increasingly misnamed Mr. Bliss. Four years later, in September of 1899, Bliss would bear one final ignominy: On that date, at the corner of West 74th Street and Central Park Avenue in New York City, Henry Hale Bliss became the country's—indeed, the Western Hemisphere's—first pedestrian traffic fatality.

As the story unfolded later, Bliss had been riding in a trolley with a somewhat mysterious female companion named Miss Lee. After alighting at his stop (at West 74th Street and Central Park West, now the site of row after row of multimillion-dollar apartments), he turned around and reached in to help his fellow passenger, at which point he was run over by an electric

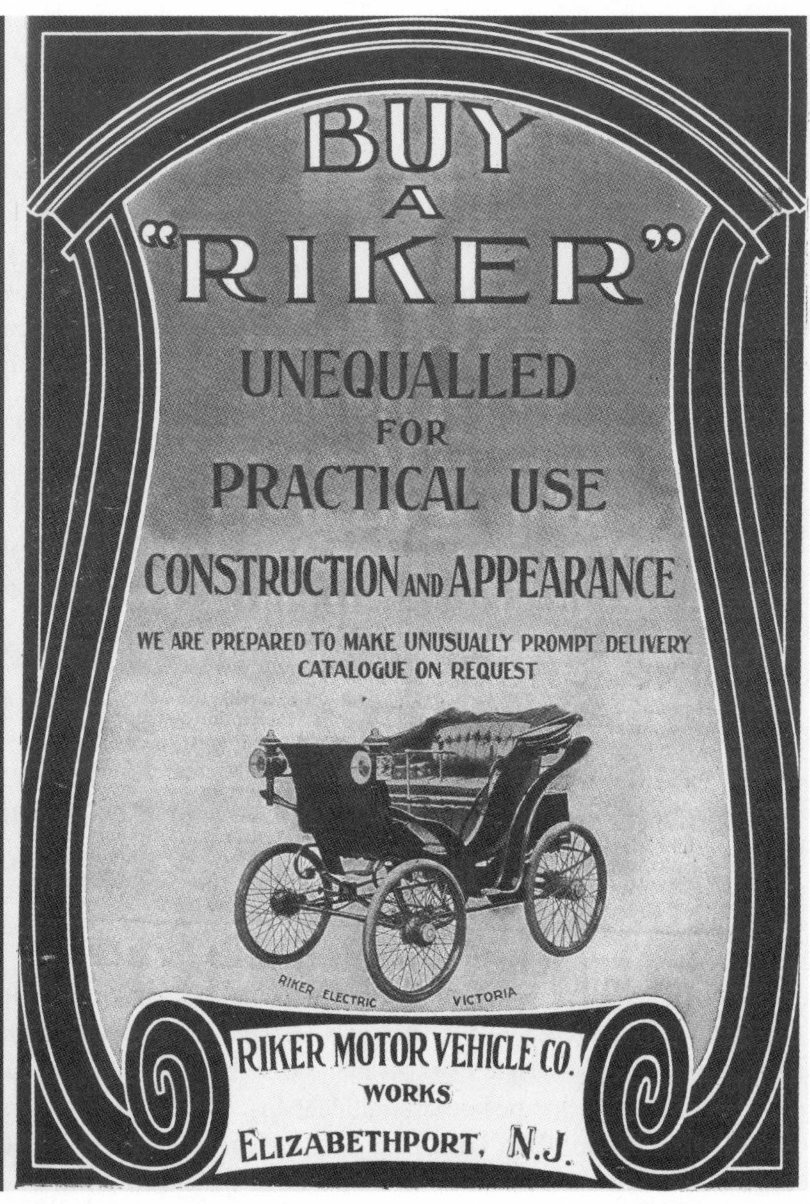

The unfortunate Mr. Bliss was probably struck by a vehicle very much like this one. The Riker Motor Vehicle Company made a couple thousand vehicles, most of which served as taxis in metropolitan areas such as New York City. *Image in the public domain.*

taxicab driven by one Arthur Smith. (Riding in the taxi was David Edson, a medical doctor and the son of former New York City mayor Franklin Edson. The younger Edson attempted to deliver first aid, but Bliss's "head and chest had been crushed," as the city's newspapers enthusiastically recounted; he died the next day. Smith was charged with manslaughter, but like Mary Alice, he was acquitted, the jury finding the death accidental and Smith thus not at fault.)

Exactly one hundred years later, on September 13, 1999, a plaque was erected at the site, both to commemorate Bliss's unfortunate end and to encourage today's drivers to take care on the road. It reads:

> *Here at West 74th Street and Central Park West, Henry H. Bliss dismounted from a streetcar and was struck and knocked unconscious by an automobile on the evening of September 13, 1899. When Mr. Bliss, a New York real estate man, died the next morning from his injuries, he became the first recorded motor vehicle fatality in the Western Hemisphere. This sign was erected to remember Mr. Bliss on the centennial of his untimely death and to promote safety on our streets and highways.*

The plaque may have served to honor Henry Bliss's memory (his great-granddaughter attended the ceremony, laying a bouquet of roses at the site of the accident), but it certainly didn't have any effect on the number of pedestrian fatalities or on the care exercised by drivers—in New York or anywhere else. These days, New York City generally sees an average of around fifteen thousand pedestrian–motor vehicle accidents and at least one hundred fatalities per year. (Across the United States, the Center for Disease Control and Prevention estimated the cost of motor vehicle accidents in 2000 at $230 billion, while in 2003 the *global* economic cost of motor vehicle accidents was estimated at about $518 billion per year.)

As we well know, automotive technology, though indisputably beneficial, can be quite dangerous. In this it is not unlike all technologies, which can be misused or abused, or which can have consequences (positive and negative) far beyond those intended or anticipated. This particular technology, no doubt seen as a boon to most, and perhaps as an exciting sign of significant new changes yet to come, had just caused its first pedestrian fatality.

The Birth of the Automobile

So, where and when did it start, this now-ubiquitous technology—or, more accurately, this collection of technologies? Who invented the automobile?

Well, it wasn't Henry Ford, in spite of how closely we associate his name with the birth of the industry. In fact, it's difficult to single out one person as the inventor; instead, several people around the world were experimenting over many decades with techniques and tools that would ultimately lead to the invention of the automobile.

We tend to think of the automobile as somehow uniquely American, perhaps because of our fascination with and love for it—and because of the great automobile manufacturing centers in and around Detroit, Michigan—but the automobile was actually born in Europe. Perhaps one place that its conception may be said to have begun was in tiny Belgium. In 1858, a thirty-six-year-old Belgian engineer and inventor named Jean Joseph Étienne Lenoir created the first commercially successful internal combustion engine. It was a single-cylinder, two-stroke engine that ran on coal gas. It was noisy and not terribly reliable, but it was a working internal combustion engine used to power a range of vehicles, most notably a three-wheeled carriage. (A later model burned petroleum and utilized a carburetor, making it an even more obvious progenitor of today's gasoline engines.)

In 1886, German engineer Karl Benz created the Benz Patent Motor Car, which utilized a single-cylinder, four-stroke engine mounted in the rear of a three-wheeled vehicle. He was awarded a patent for his efforts that many regard as—to quote one author—"the birth certificate of the automobile." Though others had preceded him, Benz's efforts, and his creation of a more robust and sophisticated vehicle, laid the foundation for the industry—and the product—we all know. It was, arguably, the birth of the modern automobile.

Benz would, of course, eventually go on to found Daimler-Benz and then, via a 1926 merger, Mercedes-Benz. (Ironically, Benz's greatest obstacles were not technical but financial, according to Michael Bock, Daimler AG's former director of Mercedes-Benz Classic and Customer Centre. "Benz first had to convince his business partners and investors of his idea to create something like the automobile," says Bock. "At that time Benz built engines, and at first his investors did not see any reason to change that.")

Karl Benz is generally credited with the invention of the commercially viable modern automobile, though many others before and after him contributed a great deal to the development of the industry. This is a replica of Benz's 1886 motorwagen. *Image licensed under the Creative Commons Attribution-Share Alike 3.0 Unported license. Photo by user and Wikipedia sysop Chris 73.*

In 1890, two French machinists, René Panhard and Emile Levassor, set up what was probably the world's first automobile manufacturing plant. Their first car was built in 1890, using an engine built by German engineer Gottlieb Daimler. (Daimler had previously worked for Nikolaus Otto, who in 1876 had invented a four-stroke engine.) Daimler Motors Corporation would eventually merge with Karl Benz's company, becoming Daimler-Benz AG, now known as Daimler AG. (It wasn't until 1926 that the first automobile designated as a Mercedes-Benz was released. Note that the Daimler Motor Company Limited was actually a British automobile manufacturer that had licensed the Daimler name beginning in 1896.)

The United States Applies Its Manufacturing Muscle

So, the pedigree of the automobile is undeniably European, although that doesn't mean that there were no American inventors working in the automotive arena. For example, brothers Frank and Charles E. Duryea, of Springfield, Massachusetts, designed a successful American gasoline-powered automobile in 1893, and won the first American automobile race, held in 1895. (The race followed a fifty-four-mile route from Chicago to Evanston and back, which the Duryeas' winning entry completed at the undoubtedly invigorating average speed of 7 miles per hour.)

In fact, by 1899, there were some thirty American automobile manufacturers, and in the years to follow, many thousands of automobiles were sold. For example, Studebaker, which had found success building wagons and carriages beginning in the 1850s, began building electric cars in 1902, switching to gasoline-powered vehicles in 1904. The Buick Motor Company was incorporated in 1903 in Detroit, Michigan, though the company was soon purchased by James H. Whiting and moved about seventy miles north, to Flint, Michigan.

By the early 1900s, there were quite a number of automobiles being produced in the United States and in Europe, but they remained rich men's toys. A top-of-the-line 1910 Oldsmobile retailed for $4,600, an amount greater than the cost of a small home. The 1903 Stevens-Duryea (a company formed after the Duryea brothers' acrimonious split) cost some $1,300, a good deal more than an average worker—or even many reasonably affluent professionals—could afford. The first Studebaker-Garford internal combustion model, delivered in July of 1904, cost $2,000. A new 1904 Daimler (the British variant) went for the equivalent of about $1,400 in US dollars, which an ad in one publication of the time characterized as "a moderate price." The 1905 Buick Model B cost $950, at a time when the average US salary stood at somewhere between $200 and $400 per year. (Skilled professionals earned considerably more, of course; it was the country's doctors, bankers, lawyers, and engineers who could afford the Model B and its automotive brethren.) The average factory worker would not be buying an automobile; they remained luxury items not meant for the masses.

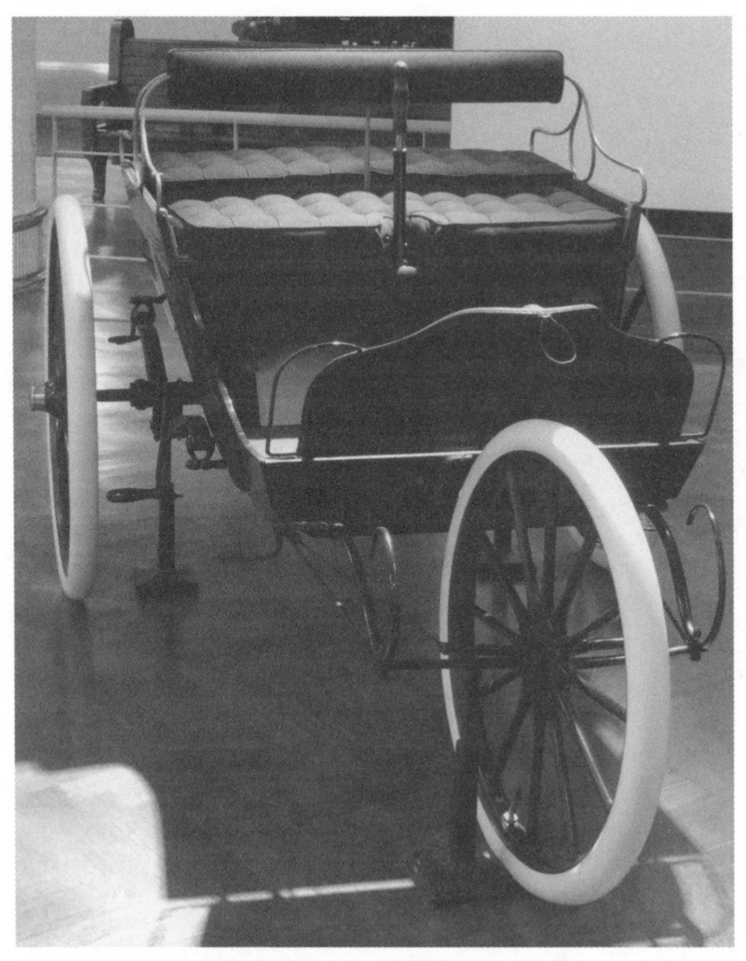

This 1899 Duryea is on display at the Henry Ford Museum in Dearborn, Michigan. Many of the earliest automobiles were three-wheeled vehicles. *Image in the public domain.*

Ford's Model T

At least, until 1908. That's the year that Henry Ford introduced the Model T. ("Any customer can have a car painted any color that he wants," he *may* have said of his new offering, "so long as it is black." Oddly enough, we don't know for sure that Ford ever actually said this. He did, in his autobiography, *write* that he had said it. Odder still, for the first few years, the car was in

This is the Model T assembly line at the Highland Park plant, in 1914. Ford invented neither the assembly line nor the automobile, but he vastly improved both, and made his car much more affordable than the competition's. *Image courtesy of the Ford Motor Company.*

fact *not* available in black; only from the mid-1900s on were all Model Ts painted that color, possibly because the use of only one color further simplified the manufacturing process.)

Henry Ford did not invent the automobile. He didn't even invent the assembly line. (That had existed, in some form or another, for many years; Ransom Olds used it in building the car that eventually became known as the Oldsmobile.) What Ford *did* do was to modernize the assembly-line concept and make it much more efficient by utilizing conveyor belts that carried parts and assemblies from one station to another. At each station was a man (or a team of men) who attached or tightened the next part before the growing vehicle literally "moved on down the line." This is a manufacturing method that's very familiar to us now, but it was revolutionary at the time. In Ford's new manufacturing scheme, the car came to the worker, rather than the other way around, and by the end of the assembly process, a new car literally rolled off the line. It was standardized manufacturing at its best—or at least, at its most efficient.

Ford was no altruist; he created a simple, affordable car for one reason: He wished to make money. Preferably a great deal of money. And he knew that the key to making money was to commoditize his product, to produce an automobile so efficiently that he could still show a good profit, even if he sold it at a price that almost anyone could afford.

He succeeded on all counts. Not unlike Adolf Hitler some thirty years later, Ford had set out to build "the people's car"—cheap, efficient, and well-made—and he had succeeded. (Interestingly, Ford was himself something of an anti-Semite, holding that "The genius of the United States of America is Christian . . . and its destiny is to remain Christian." He purchased his hometown newspaper, which ran a series of anti-Semitic articles that were eventually bound together in volumes called *The International Jew*. Hitler in fact displayed a portrait of Ford in his office, and once told a reporter, "I regard Ford as my inspiration." On the other hand, as far as we know, Ford never advocated a "final solution" of the sort that Hitler envisioned. In fact, Ford later apologized for the articles and sold the newspaper in which they first appeared.)

"It will be large enough for the family," Ford said of his automobile, "but small enough for the individual to run and care for. It will be constructed of the best materials, by the best men to be hired, after the simplest designs that modern engineering can devise. But it will be so low in price that no man making a good salary will be unable to own one, and enjoy with his family the blessing of hours of pleasure in God's great open spaces."

Ford was as good as his word. The Model T made its first appearance in 1908 and was an instant winner. (It was so successful that production of the car continued until May of 1927.) The Model T was simple, reliable, and well made. It was the first vehicle produced on moving assembly lines, and the first to be made completely of interchangeable parts.

And most of all, it was *affordable*. Ford kept increasing the efficiency of his manufacturing methods (often to the detriment and discomfort of his workers) and lowering the price of the car until, by 1918, one could buy a brand-new Model T for as little as $345. (The price would eventually drop even further.) Ford was making a fortune, and he paid his workers the princely sum of $5 per day (well, more or less; it depends on how you add it up, and there were lots of strings attached), mainly because he needed to give them a wage that guaranteed they would stay on at what had turned into a thankless, grueling job. (There might be some truth to the long-held notion that Ford also wanted his employees to be able to afford to buy Fords of their own, but possibly not. As Adam Smith Institute Fellow and writer Tim Worstall has pointed out, that logic is flawed; imagine, he says, what would happen if Boeing had to pay each of its workers enough money to buy a new jetliner.)

In any case, people bought Fords in droves. Everyone, it seemed, was driving a Model T. In fact, by 1918, half the cars on US roads were Ford Model Ts.

What almost always happens with a new technology had happened once again: Automobiles, which had begun as custom-built playthings of the wealthy, had ended up (and remain to this day) ubiquitous machines, mass-marketed, manufactured on a large scale, and affordable to most employed people.

Ford's Legacy

In 1918, one could buy a new Model T for $345, roughly 30 percent of the typical average yearly salary for that year, which was somewhere around $1,500, according to figures from the Bureau of Labor Statistics. Compare that $345 to the $1,398 that the city of Boston spent that year on a new Buick runabout purchased for use by Thomas F. Sullivan, the manager of its public works department. (Of course, Buick had several models available in 1918; some cost more, some less.) The Buick would essentially cost the average worker his entire year's salary, instead of the 30 percent of that salary required to purchase a Ford. Which one would he buy? Which one *could* he buy?

Fast-forward to more-recent times. According to the Social Security Administration, the average US salary in 2013 was about $43,000. The MSRP for a new Ford Fiesta (to stick with perhaps the least-expensive model in the company's lineup) was between $13,000 and $18,000, meaning, you might have plunked down about $14,000 on a new Fiesta a couple of years ago. The

This restored 1922 Buick touring car is typical of Buick's offerings at the time: beautiful and superbly engineered, but more expensive than the Fords of the day. This one is displayed at the 2006 Bay State Antique Automobile Club Vintage show. *Image licensed under Creative Commons CC-BY-SA-2.5. Photo by Stephen Foskett (Wikipedia user sfoskett.)*

Ford Fiesta—true to its parsimonious roots—*also* costs something close to 30 percent of that $43,000 salary. (Keep in mind, of course, that "average" is not really a terribly good indicator of . . . well, *anything*, actually. In this case, the "average salary" numbers are terribly skewed by a small percentage of people making a much larger salary and by some who make quite a bit less. But one must make do with what one has to work with, and the 1918 figures are a bit murky, so, "average" it is. And, in the end, it does help us to arrive at a reasonably useful picture of the automotive economics at work here.)

Thus, the economic revolution wrought by Ford and his Model T seems to have remained in effect: It's still reasonable for a US worker to expect to be able to purchase a new (and admittedly low-end) automobile for about

30 percent of one's salary—assuming that one's salary approaches or exceeds "average." A technology that was once beyond the reach of the common man and woman is now so readily available that, in an industrialized nation, when we encounter someone who does not own a car, we either assume a certain level of poverty, or else a conscious decision occasioned by either geography or sociopolitical inclination: Many big-city residents who could afford an automobile instead choose not to have one because it's not worth the trouble, given access to mass transit and the expense and hassle of parking, commuting, garaging, etc. Meanwhile, some simply opt out of participating in what they feel is a wasteful, environmentally unsound enterprise. But only rarely is someone in a first-world country without an automobile for financial reasons alone. (About 90 percent or so of American households own a car, though the number of carless households has risen a little over the past few years, as more people carpool, bike, take advantage of mass transit, or even—gasp!—*walk* from one place to another.)

It's a Brand-New World

But Henry Ford—thanks in large part to the automotive pioneers that preceded him and made his success possible—did more than transform the nascent automobile industry; he also revolutionized a country, and perhaps the world. (Interestingly, Ford did so at a relatively advanced age. "Ford was a two-time failure in the automotive industry," notes Matt Anderson, curator of transportation at The Edison Institute, a Dearborn, Michigan, museum founded by Ford in 1929. "Neither of his first two companies survived under his leadership. Ford was not the typical 'young and hungry' entrepreneur; he was nearly forty years old when he established Ford Motor Company, with a family to support. He was fifty when he implemented the moving assembly line and the Five Dollar Day. Henry Ford is a case study in persistence, and a reminder that success sometimes comes later in life," says Anderson.

Ford even changed the physical geography of the country. In the United States, what had been barren stretches of hills, mountains, and pastures were soon riven by country roads and then bisected by national highways. Gas stations appeared, both in cities and along lonely stretches of road. (In the parlance of the era, these were "service stations," and one could indeed find real service there—complete with uniformed attendants eager to wash your windshield, check your oil and water, and more, in addition to filling your tank with gasoline.)

"It is fair to say that the Model T revolutionized American life," says Kenneth C. Davis, in *Don't Know Much About History*. "When Congress enacted highway fund legislation in 1916 and the country embarked on a massive road-building era, the American dream of freedom on the open road became a reality. In a short time, the auto industry became the keystone of the American economy, in good times and bad."

The arrival of the automobile helped to create, as successful technologies are wont to do, dozens of spin-off industries. It ultimately led to millions of jobs, not just in the auto industry, but also in allied—and often in seemingly unrelated—industries. The price of vulcanized rubber skyrocketed, demand for steel shot up, massive road-construction projects became commonplace. (The granddaddy of all of those, the 47,000-mile US Interstate Highway System—officially, the Dwight D. Eisenhower National System of Interstate

This is the map of the US Interstate Highway System as it currently exists. Begun in the mid-1950s, the system comprises about 47,000 miles of roadway and is still not completely finished. *Image released into the public domain by its author, SPUI.*

and Defense Highways—begun in 1956, employed thousands of people, and cost the federal government about $120 billion.) And where the automobile went, motels ("motor hotels") also went, appearing along major roads and highways. Roadside diners sprang up. Too, someone had to do bodywork, and upholstery, and engine repair. Someone had to produce (and warehouse and ship) literally tons of glass to go into the windows in manufacturers' vehicles—and to replace those windows that were damaged in use. And on and on it went, with hundreds of other industries affected—or even created—by the expanding automobile industry.

There were also far-reaching social effects, and surely Ford had not foreseen these. People began to choose to live farther from work. Young people, especially, took advantage of the freedom granted by access to the automobile. (Some have theorized that the nascent "sexual revolution" originated during the time that young people spent on the road—or parked just off of it—finally away from home and unchaperoned. There may be something to the idea, but then it seems that *every* generation, when young, is intent on creating a sexual revolution of one sort or another. That is part of what it means to be young, after all.)

As people discovered that they needn't live in the cities, or in fact anywhere near their places of employment, suburbs took root, leading to the idea of "bedroom communities." This led ultimately to cookie-cutter tract developments such as Levittown, built on a scale and in a fashion that deliberately mimicked mass-manufacturing methods similar to Ford's.

Bill Levitt, a former Seabee, had come to appreciate fast, efficient construction methods during World War II, and when he began building his (at the time, much-disparaged) developments, he brought to those efforts an assembly-line sensibility, but reversed. In Levitt's case, the workers moved from one house to the next, each doing one specific task: One man, for instance, would work his way through all of the houses, installing windows. Another would install a picket fence. (The *same* windows and the same picket fence, it should be noted.) And so on down the line, in a precisely choreographed ballet of cheap, efficient fabrication that almost amounted to something more akin to *assembly* than to traditional construction.

This approach allowed Levitt to avoid hiring expensive craftsmen, and made the typically high turnover rate more palatable. The result was a collection of houses that looked essentially alike and which lacked the elegance (and perhaps the sturdiness) provided by the old-world craftsmen who had hand-built the custom homes of an earlier age; these, indeed, were the very "failings" that the critics seized upon. But, just as Ford's Model T had enabled the average worker to purchase an automobile, the Levittown communities (there were eventually several of them, and they truly were designed to be communities) and subsequent tract developments provided affordable housing to a nation of young people just returned from a horrific war, and determined to own a small piece of the American dream for which they had sacrificed so much.

The social and geographic connections between Ford and his Model T and Bill Levitt and his suburban developments are unmistakable, tangible, and still visible today. The numbers tell the tale: In 1900, there were fewer than ten thousand passenger cars in existence in the entire world. By 1920, there were ten million. By the end of the century, there were half a billion. Over the same period of time, and not at all coincidentally, the share of the US population that lived in suburbs doubled between 1900 and 1950, and

A dapper Henry Ford poses in 1919 or thereabouts. He would have been about 56 years old at the time. *Image in the public domain.*

then redoubled between 1950 and 2000, by which time 52 percent of the country's population lived in the suburbs. (And to underscore the connection even further, consider that many of those living in the suburbs actually work in a major metropolitan area, or in a different suburb. And, notwithstanding a recent push to improve the availability and efficiency of mass transit, the vast majority of those workers commute by automobile—and most of them drive alone, meaning that even more autos are on the road.)

Many contributed to building it, but one man revolutionized the auto industry.

Another man transformed an entirely different industry, but it's quite possible that without Ford's Model T, there might have been no Levittown, and perhaps no suburbs as we know them now. Most likely unforeseen by Ford, a car—*his* car—changed the world.

Tech innovators never know, really, what the ultimate effects of their labors might be. Henry Ford certainly never foresaw the far-reaching results of his work.

"I do not think that Henry Ford fully understood the way in which the automobile would physically and socially reshape the nation," says Matt Anderson of The Edison Institute. "He obviously saw the utility in the auto—as something more practical than a leisure toy for the wealthy—but he saw it as a tool with which to end rural isolation, and perhaps as a way to tie together the rural farm and the urban factory. I don't think there was any way that he could have conceived the shape and scale that that freedom would reach a century later. He certainly wasn't thinking about interstate highways, shopping malls, or fast food when he introduced the Model T. Ford didn't die until 1947, so he saw some of these changes take hold, but he did not survive to see the post–World War II boom that really established our 'car culture' as we know it today."

Of course, these days we don't travel only by car. In order to get from one place to another, we also utilize trains, ships, and that most romantic and technologically impressive of all conveyances, the airplane. We're still waiting for the true "Henry Ford of aviation" (William T. Piper of Piper Aircraft has in fact been called that) to appear and create the Model T of aircraft, but that could happen, perhaps in our lifetime. We'll look at some of the technology associated with flight in the next chapter.

Chapter Five

Drones: Robots in the Skies

Technology is so much fun but we can drown in our technology. The fog of information can drive out knowledge.

—Daniel J. Boorstin

First Flight

These days we're used to hearing about companies, large and small, "going public." Facebook, for instance, famously went public in 2012, selling shares in an IPO (initial public offering) and quickly raising some $10 billion, putting the theoretical value of the company at well over $100 billion. This made Facebook's IPO one of the largest in history. (Ironically, the company's share values were not as high as had been expected, and the IPO was something of a disappointment, in spite of the sums that were raised. For what it's worth—and it's apparently worth quite a bit—falling prices over the first week or two were more than offset by later gains; by December of 2015, the company was valued at over $230 billion.)

Companies "go public" when they move from being privately held to being in effect owned by the public—or at least that portion of the public that is willing to buy those initial shares. The IPO is a time-honored mechanism used to raise funds, pay off early investors, and provide an exit strategy for any venture capitalists who might have provided seed funds.

How time-honored? You have to go back to the fall of 1909 to find the first public company in the United States. President William Howard Taft (he of the enormous girth and equally enormous appetites) was in office.

No. 821,393. PATENTED MAY 22, 1906.

O. & W. WRIGHT.

FLYING MACHINE.

APPLICATION FILED MAR. 23, 1903.

3 SHEETS—SHEET 2.

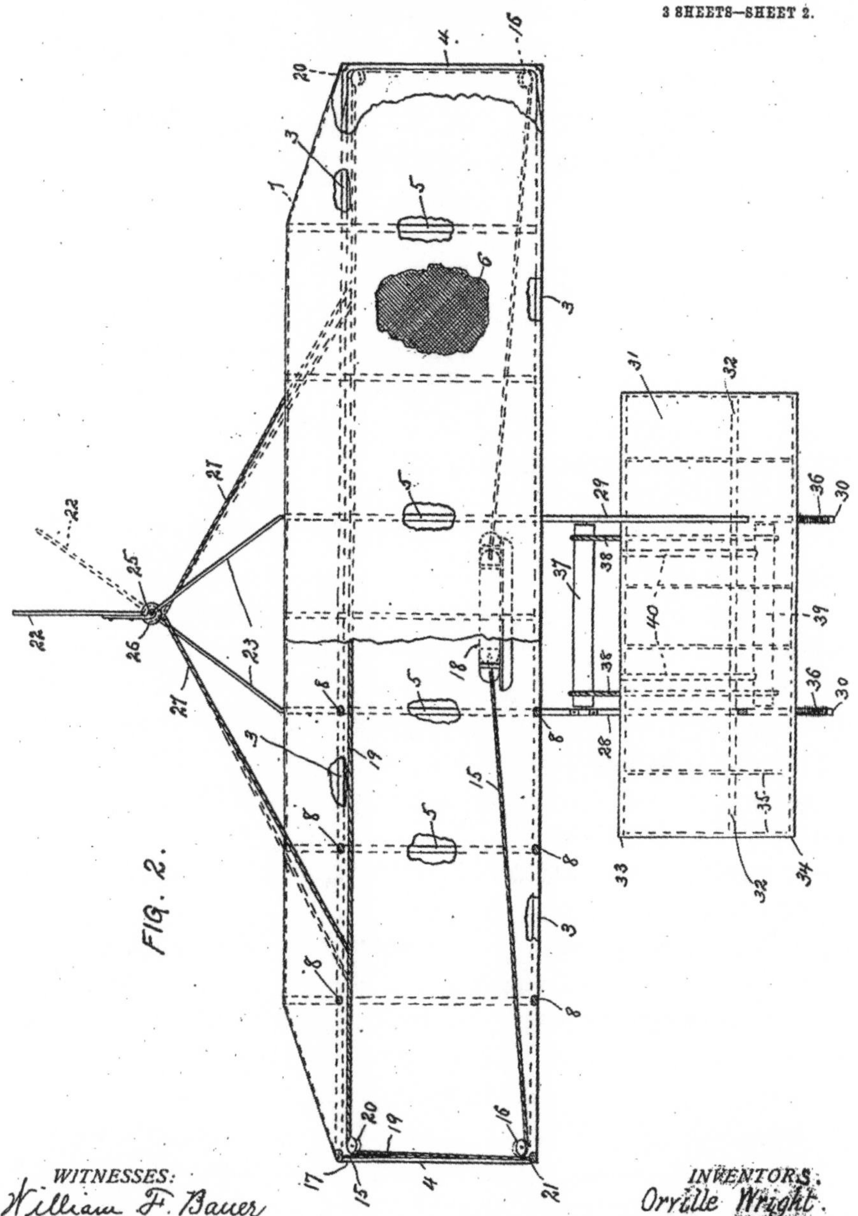

FIG. 2.

WITNESSES:

William F. Bauer

Irvine Miller

INVENTORS.

Orville Wright.

Wilbur Wright.

BY H. A. Toulmin.

ATTORNEY.

This drawing was submitted by the Wright Brothers' attorneys as part of their initial patent application. *Image courtesy of the US Patent Office.*

Leo Fender (who, oddly enough, never did learn to play any of the guitars he built) was born near Anaheim, California. The last of the troops stationed in Cuba during the Spanish-American War were about to leave. And the United States Army bought the world's first military airplane, spending $30,000 on a Wright Military Flyer Model A that was capable of cruising at a breathtaking 40 miles per hour.

And that's where our story truly begins. That first public entity was none other than the Wright Company, formed only six years after the famed brothers' first powered flight in December of 1903. The company, capitalized at $1 million, aimed to manufacture and sell "aeroplanes" to the military and to individual pilots. (Interestingly, while the brothers were brilliant inventors and engineers, they were somewhat less effective as businessmen, and were not nearly as financially successful as they might have been, though in the end they did each manage to amass sizable fortunes.)

Note that some sources (including www.wright-brothers.org) say that the brothers' first US patent (#821,393) did not claim invention of a flying machine, but rather the invention of a system of aerodynamic control that manipulated a flying machine's surfaces. The actual patent description, however, certainly seems to describe not merely a control system, but the entire "aeroplane." In any case, the patent—and the brothers—were not without their detractors, and the brothers' claim to the invention of the airplane was the subject of much debate.

The Langley Aerodrome

Some argue that, even before the Wright Brothers' success, astronomer/inventor (and eventually secretary of the Smithsonian Institution) Samuel P. Langley may have succeeded in powered flight—depending on how we define both "powered" and "flight."

Langley began experimenting with flight in the late 1880s. Using rubber bands for power, he built models that flew for several seconds and covered distances of up to one hundred feet.

From there Langley moved on to steam power, and by 1893 was ready to unveil what he called Aerodrome No. 4. The unmanned steam-powered device was meant to launch from the roof of a houseboat moored in the Potomac River, which sounds silly until you realize that it was anything but; the use of a boat allowed Langley and his team to turn the boat such that the Aerodrome could be launched into the wind, regardless of the direction of that wind.

The flight of Aerodrome No. 4 was, however, beset by problems, including the fact that, when launched, the aircraft didn't actually fly so much as *fall*. It wasn't until October of 1894 that Langley's pilotless "aeroplane" succeeded in making a short hop (about 130 feet) over the river, and in December of that year, a newly constructed Aerodrome No. 5 flew about one hundred feet. After that, things started going well for Langley. In 1896, Aerodrome No. 5 managed two flights of well over two thousand feet, and in November of that year actually flew over four thousand feet, staying aloft for more than one minute, but all with no pilot aboard.

In October of 1903, Langley was, apparently at the behest of President William McKinley (and given by the government a then-substantial $50,000 budget with which to work), experimenting with the (manned) Aerodrome A, which sported a 52-horsepower, five-cylinder radial engine that would not look completely out of place on a propeller-driven aircraft of today. However, the first launch was a disaster, with the craft plunging ignominiously nose-first into the Potomac River. Langley and his team regrouped. (The Langley team was getting quite good at regrouping.)

In December, they tried again, and the failure was, if anything, even more spectacular. The rebuilt plane reared up, pointing almost vertically at the sky, and then fell back into the river once again. Pilot Charles M. Manly, who had helped design the craft's radial engine, was briefly trapped beneath the wreckage, but was rescued unharmed, possibly thanks to the cork life vest he was wearing.

A few days later, on December 17, at Kitty Hawk, North Carolina, Wilbur and Orville Wright's Wright Flyer, piloted by Orville, would flit a few feet above the sandy beach, landing uneventfully about twelve seconds after takeoff some 120 feet from the point at which it took off.

Thus ended Langley's aeronautical experiments, and the flight fraternity immediately entered upon a debate that rages until this day. Did Langley's aircraft "fly"? The Smithsonian (of which Langley was, after all, the secretary) thought so, and made its opinion known, which set off a long-running feud with the Wright brothers. (Eventually the museum relented, more or less: It adorned the Langley exhibit with a plaque noting that the plane was *capable* of flight, thus neatly skirting the troublesome twin issues of whether it *had* actually flown, and whether having a pilot aboard was a significant consideration.)

A Pair of Bicycle Mechanics Change the World

In spite of Langley's efforts, and he was surely a brilliant scientist and a dogged experimenter, Orville and Wilbur Wright, two bicycle mechanics living in Dayton, Ohio, would be the ones to gain fame as the inventors of manned heavier-than-air flight. The brothers were nothing if not dedicated, and, even though they were only amateur scientists, they thoroughly researched all the existing literature on heavier-than-air flight. They spent years designing and flying gliders, working with them until they understood the mechanics of flight, even going so far as to build a rudimentary wind tunnel to use in testing their designs.

(The wind tunnel was used mainly for testing various wing surface shapes. The brothers were serious and persistent investigators. Early on in their research on the mechanics of flight, Wilbur had written to the Smithsonian Institution "to obtain such papers as the Smithsonian Institution has published on this subject, and if possible a list of other works in print in the English language." Thankfully for the sake of the future of aviation, an assistant secretary of the Smithsonian responded, sending Wilbur a list of books and a supply of pamphlets.)

Understanding as no one had before them that what was needed to enable heavier-than-air flight was not an ever more powerful engine (a research path that led some other experimenters astray), but instead, adequate control surfaces to regulate pitch, yaw, and roll, the brothers nonetheless knew that they *did* need a reliable lightweight gasoline engine that could generate enough power to turn the two handmade "pusher" propellers on their Wright Flyer I. (Pusher props sit behind the pilot, *pushing* the plane forward, while tractor props are located at the front of the plane, *pulling* the plane forward. In the Wright Brothers' case, the two propellers turned in opposite directions so as to cancel out each other's rotational torque.)

They couldn't find an engine that would do the job. Engines were cast mainly of iron at that point, or of alloys that, while strong enough to withstand the combustive forces at work, were all much too heavy.

The brothers found the answer to their problem right in their own shop: Mechanic Charlie Taylor, a Nebraska transplant who had ended up in

Dayton, Ohio, after his marriage, had gone to work at the Wright Brothers' bicycle shop in June of 1901. Paid a princely salary of $18 per week, it was Taylor who had built the brothers' wind tunnel and machined many of the parts for their gliders. After exhausting their attempts to find someone who could provide a reliable, vibration-free engine that would be light enough to mount in their airplane, the Wrights asked Charlie to build one.

Charlie, though he had never before built a gasoline engine, and working with no real plans, produced a lightweight aluminum-block engine in only six weeks. Using a powered lathe, Taylor machined almost every part of the new engine, including the connecting rods, intake valves, exhaust valves, pistons, valve guides, and rocker arm.

The completed engine, a 152-pound four-stroke without a carburetor, eventually generated 12 horsepower, enough power to allow the brothers to add another 150 pounds of struts and braces to the design of the Wright Flyer.

The rest is, quite literally, history. After an abortive attempt a few days earlier, on December 17, 1903, the Wright Flyer flew 120 feet, staying in the air for about twelve seconds, only a few feet above the sands at Kitty Hawk, North Carolina. Manned, controllable flight was a reality; Charlie Taylor's little engine, combined with the brothers' innovative control mechanisms, eventually led to a reliable airplane.

Flight Comes to (Some of) the People

We know what happened after that. A war (and after the war, devoted barnstormers and experimenters) pushed the technology forward. Advance led to advance—as is the way with technology—and soon, heavier-than-air flight was a common occurrence.

Common, but expensive. We've seen that the modified Wright Flyer was sold to the military for US $30,000, this at a time when the average unskilled laborer in Ohio, where the Wright brothers lived, was making about $325 per year, and a skilled craftsman might make, at most, $1,000 per year. Even if the military price were vastly inflated (a not-uncommon practice, even then), the typical wage earner could never hope to come up with anything close to that amount. About seventeen years later, the custom-built plane in which Charles Lindbergh crossed the Atlantic was said to have cost about US $10,500 in 1927, or about $144,000 in today's US dollars.

By the early 1930s, aircraft looked more like this 1933 Flagg biplane. *Public domain image courtesy of Tim Vickers. Image released into the public domain by the photographer, Tim Vickers.*

For the time being, there would be no "citizen pilots" owning their own planes, but that was—and is—changing.

Aircraft are still expensive, of course, but the proportion of the average salary required to purchase one has dropped dramatically. In 2013, the per capita income in the United States, for instance, was about $53,000. In 1927, the per capita income in the United States was about $2,000, according to the National Bureau of Economic Research. That means that if you lived in the United States in 1927, it would take about five years' worth of salary to purchase an airplane such as the one Lindbergh flew to France. These days, purchasing a used $100,000 airplane would require about two years' worth of salary. (In fact, today I can purchase an older but still serviceable single-engine prop-driven airplane for anywhere between US $20,000 and US $100,000. Although frankly, I would be reluctant—read: terrified—to trust my life to a $20,000 airplane. Then again, I'm not terribly comfortable trusting my life to a $20 million airplane.)

Of course, that's assuming you were to purchase a used plane; Lindbergh's *Spirit of St. Louis* was not only brand-new, it was also custom-built. But these days there are, in fact, plenty of used airplanes to purchase, which may not have been the case in 1927—or even in 1950. (At press time, one of the USA's premier aircraft listing services, Controller.com, lists over 1,700 used single-engine prop-driven aircraft for sale; these range in price from $20,000 or $30,000 all the way up to $1.3 million. The vast majority, though, will sell for $150,000 or less. Still quite an expensive undertaking, but much less so than in years past. And if you're really brave, you can find a twin-engined Cessna very much like the one Sky King flew in the 1950s television show for less than $50,000. Just don't ask me to fly in it with you.)

Enter the Drones: Bugs and Bats

Even as aircraft became more and more sophisticated, there was one piece of equipment that could always be counted on and which could never be replaced: the pilot.

Until 1917, anyway. That's when the Dayton-Wright Airplane Company (yes, Wilbur and Orville again) built the Kettering Aerial Torpedo, aka, "the bug." It wasn't quite what we'd think of as a drone, but it was an obvious precursor, being pilotless and guided by preset controls. After the appropriate predetermined amount of flight time had passed, the engine shut down, the wings came off, and suddenly, the "bug" was no longer an airplane at all—it was now a bomb, one that carried about 180 pounds of explosive.

Later, as the technology was pushed forward by yet another war, the military experimented with a radio-controlled gliding bomb known as "the bat." Then, in the mid-1940s, a gliding bomb known as the GB-1 was developed. This was essentially a bomb around which a small radio-controlled glider had been built. These were very accurate and quite lethal. The GB-1 was followed by the GB-4, which was the first "television-guided" weapon; it was not nearly as effective as its theoretically less-sophisticated forerunner, though, because the television images of the time were so crude that the GB-4 could function only if the weather were perfect.

The technology took another leap forward in the 1950s, when the Ryan Aeronautical Company developed the Firebee. This was a jet-propelled drone that could stay aloft for two hours, flying at sixty thousand feet at speeds of up to 500 knots. For several years, these drones served as test tar-

Initially used as practice targets, the Teledyne Ryan Firebee UAV ushered in the era of modern drones. *Licensed under the Creative Commons under the terms of the GNU Free Documentation License. Photo courtesy of user Bukvoed.*

gets for pilot training and weapons testing, but in 1971, a modified Firebee took to the offensive, scoring several (simulated) "hits" on an F-4 Phantom in flight. The next year, a newer Firebee scored a (simulated) direct hit on the navy destroyer *Wainwright*, in spite of the fact that the ship was protected by a battery of missiles. The military potential of drones, as both offensive and defensive weapons, was now a matter of record.

Drones had been used successfully for reconnaissance, photography, and other intelligence-gathering operations all through the 1960s, but the times, they were a-changing, as Dylan had pointed out. And drones were about to change radically, largely because of an émigré from Israel named Abraham Karem. Born in Baghdad, Karem's family moved to Israel in 1951, and Karem himself later emigrated from Israel to Los Angeles. Once settled in, the resourceful aeronautical engineer went into his garage and began building. One year later, the Albatross emerged.

The Albatross was meant to serve as a demonstration of what could be accomplished with military UAVs (Unmanned Aerial Vehicles), and it did its job exceedingly well. Weighing only two hundred pounds and carrying a television camera in its nose, the slim craft could stay aloft for an unimaginable fifty-six hours. After DARPA (the Defense Advanced Research Projects Agency, the DOD's research arm, eventually responsible for building what would someday become the Internet) realized what Karem and his team could contribute, the organization funded the development of Amber, the next drone to emerge from Karem's little company, now called Leading Systems. Amber was radio-controlled and could either take off and land like a conventional aircraft, or its wings could be folded such that the drone could be fired by rocket from a canister. In addition, the drone could land almost vertically, making it ideal for retrieval by a small ship, a truck, a submarine, or a trailer.

Most importantly, Amber was reliable. With conventional aircraft, there was always a pilot on board to make quick decisions requiring experience, good judgment, and . . . well, some sort of flier's intuition. A drone, even one controlled by an operator on the ground or in another aircraft, doesn't have the advantage of a pilot right there to take over when things start to get out of hand. Amber's performance, however, was almost flawless: It was flown for 650 hours without the loss of an aircraft. Compared to other drones, that was an amazing record.

The military drones with which we're all familiar, such as the Predator and its MQ-9 Reaper variant, are direct descendants of Amber. These newer drones cost tens of millions of dollars apiece, and can stay aloft for more than forty hours while carrying a thousand pounds of munitions or other payload.

Obviously, these aircraft are nothing like the little Kettering Aerial Torpedo, the "bug" built in 1917.

Now, municipal agencies, including local and regional law enforcement agencies, are getting into the act.

In 2011, for instance, the Montgomery County, Texas, sheriff's office used a Department of Homeland Security grant to acquire a ShadowHawk Unmanned Aerial System, a sophisticated helicopter drone meant to provide safe and efficient aerial surveillance and to aid the department's SWAT team in its search-and-rescue operations. A few years after it was acquired, and after only a few actual deployments, operators managed to crash the $250,000 machine into Lake Conroe, about sixty miles outside of Houston. The Houston Police Department's dive team was called in to retrieve the drone, pieces of which were recovered about forty feet beneath the surface of the lake. (The HPD called off its own plans to deploy "traffic ticketing" drones after a Houston television station owned by the *Washington Post* broadcast a story that showed local officials attending a "secret" drone air show some seventy miles outside of Houston.)

A Drone of Your Own

These sorts of drones are certainly nothing any of us could afford. (Which is undoubtedly a good thing, considering the devastation a modern, weaponized drone can wreak. That much firepower in the hands of the untrained, the unsupervised, and the very possibly unsound, would be a terrible thing to consider.) But, seemingly all at once, you and I *can* nonetheless own and operate some fairly sophisticated (but still relatively affordable) drones. And so can companies ranging from Hollywood production studios to real estate sales groups, and from search-and-rescue teams to surveyors. Drones of various sizes and capabilities can be had for as little as a few hundred dollars; for a few thousand dollars, one can acquire an amazingly sophisticated UAV. And in between are dozens of UAV options, with more arriving all the time and prices continuing to drop.

Take the popular drones produced by DJI, a Chinese company with offices in Los Angeles, California; Schondra, Germany; and Osaka, Japan.

"We're only now seeing the possible uses for affordable commercial and personal drones," says DJI spokesman Michael Perry. "Everyone from archaeologists to construction workers, search-and-rescue teams, firefighters, filmmakers, architects, sports teams, and farmers (the list grows every week) has found tremendous benefits from being able to get a new perspective on their work."

Perry notes that although consumer-grade drones will surely increase in popularity, "the largest growth will likely happen in the industrial applications for the technology."

The truth is, we can't even begin to see the possible uses for civilian drones, nor can we imagine the ramifications of those uses.

"As a real estate agent, my wife is interested in the potential for bird's-eye shots," says technologist Chris Angelini, the former editor in chief of the Tom's Hardware technology website. Angelini, who recently built his own drone, a $3,000 Team BlackSheep Discovery Pro, also notes that he's heard a lot about drone uses in the oil business (the drones often armed

Technology writer Chris Angelini is a drone aficionado with whom we spoke about some of the challenges of drone technology. He recently built a $3,000 Team BlackSheep Discovery Pro drone of his own. *Image courtesy of Chris Angelini.*

with infrared cameras, in that case) and in wedding photography. (If you're interested in following along, Chris documented his experience building the drone at the Tom's Hardware website.)

Some potential uses are still beyond most consumer-level drones, of course, but the equipment is improving.

"Right now," says Angelini, "I'm looking at [a flight time of] eight or nine minutes, max, from a charge. At some point, I want to see commercial craft able to stay in the air for an hour or more at a time, with programmable waypoints and the intelligence to log or report specific events. A farmer could use something like that to survey acres of land. An energy company could use that to control the environmental impact of an oil field. The forest service could patrol remote areas for illicit drug grows."

Practical and Beneficial Uses

Even at this embryonic stage, though, drones have proved useful in research, search and rescue, firefighting, farming, law enforcement, and other areas.

In the tornado-prone Midwest, drones are being used in a project being conducted by University of Colorado and University of Nebraska researchers to track and better understand these destructive storm systems. The specialized drones can fly at 2,500 feet, higher than ground-station instruments can see, and they can also fly as low as 300 feet, which is lower than radar can reach.

Drones being flown in or near super-cell thunderstorms of the type that spawn tornadoes could be destroyed or damaged, of course, but even at about $30,000 to $50,000 each, that's significantly less than the cost of losing an airplane—to say nothing of losing the airplane's pilot.

DJI's Michael Perry points to a wide variety of beneficial uses for drone technology.

"Relief workers in the Philippines used our S800 [drone] to survey areas hit by Typhoon Haiyan to better focus relief efforts," he says. "In Wisconsin, a drone pilot found within two hours an elderly man who had been missing for three days. And a Connecticut firefighter recently used his [DJI] Phantom to fly over a burning building to monitor the spread of the fire towards an inaccessible, potentially explosive storage area."

But that's not all. Beneficial uses of the technology continue to spring up, sometimes from the most unlikely places. The Peruvian government is using the DJI Phantom drone to create 3D maps of Incan ruins to ensure that land development doesn't destroy their cultural heritage. The Nature Conservancy flies DJI F450 drones over bird populations at night to monitor their migratory patterns. Journalists in El Salvador used a Phantom drone to report live from voting stations during the country's recent presidential elections, and the BBC used drones flying over protests in Thailand to show the scale of the unrest in Bangkok.

The Phantom 2 Vision+ drone is one of DJI's most popular products. It is essentially a flying platform for a stabilized, gimbaled camera. It's controlled by a combination of a supplied dual-joystick controller and a "Ground Station" smartphone app. *Image courtesy of DJI.*

And, in a development that almost literally involves the piggybacking of one technology on another, researchers from the University of North Texas are developing antennas to mount onto DJI F550 drones to provide Wi-Fi support to victims in disaster zones. If that effort is successful, the drones themselves will become airborne hotspots that could allow people to communicate, call for help, or relay conditions to the authorities.

On the downside . . .

Privacy and Safety Issues

The rise of drones brings with it two thorny issues, the solutions to which are neither simple nor universally agreed upon.

First, drones can be used to spy on people. After all, we're talking here about what amounts to a computer-based flying camera. How many nosy neighbors do you think will use it to peer into someone's second-story window? In fact, since even inexpensive drones can easily reach altitudes of several hundred feet, there's really nothing to stop them from recording the view into a twentieth-floor hotel suite or office building, assuming that the curtains or blinds are not drawn.

Or consider for a moment the related issues of stalking, bullying, and domestic abuse. The technology is so widely available and so potentially invasive that it can't help but exacerbate what has become a nation-wide problem. Already, a couple in Seattle has accused a neighbor of drone-abetted stalking. And the state of Connecticut—although unable to regulate commercial drones, since that is the purview of the FAA—is looking into whether or not it can regulate government-owned drones, including those operated by law enforcement agencies. Meanwhile, a bipartisan effort in Wisconsin is attempting to pass drone-related laws based on Fourth Amendment protections; that is, laws centered on the Constitution's guarantee of freedom from unreasonable search and seizure of property.

However, an angry ex or a snooping neighbor are really the least of your drone-related privacy worries. Those can largely be dealt with by contacting law enforcement.

But what if the problem *is* law enforcement? More and more police departments are looking into drones,

The HEXO+ is an autonomous drone that links to your smartphone and then follows you, filming your activities. *Image courtesy of Squadrone System, Inc.*

and not just to assist in SWAT deployments. The Los Angeles Police Department, for instance—no stranger to accusations of spying, and worse—has acquired drones that it says it plans to use in situations involving hostages or barricaded suspects. Yet, as *L.A. Times* analyst Jim Newton has noted, "[T]he [LAPD] drones are just one aspect of a profound reconsideration of the relationship between policing and privacy. Especially in the area of fighting terrorism, police are moving from solving crimes to anticipating them, aided by data mining and other technologies. The new techniques carry with them the possibility of enhancing public safety, but they give some people the creeps."

It's difficult to rely on law enforcement to resolve problems when law enforcement is in fact the problem—or is at least perceived to be.

The second issue is safety, and we're not talking just about the damage an errant drone might cause if it dropped down out of the sky and landed on your head. (DJI's Phantom 2 Vision drone weighs several pounds; that weight, falling from a height of anything more than a couple of feet, can do some serious damage, and that's without taking into account the four whirling propellers.) That threat is worrisome enough, but it's not the safety issue that has legislators (and pilots) up in arms. What's worrying pilots (and no doubt more than a few passengers) is the thought of one of these machines being sucked into a jet engine as an aircraft is taking off or is on approach; the result could be horrendous. Having a cool, techie hobby is one thing; killing a few hundred people just because you're flying too high near an airport is a (potential) tragedy of the highest order.

The result is that governments (in the case of the United States, the FAA) have had to come up with enforceable rules governing who can fly even smaller drones, as well as how and where they can be flown. If that means requiring a flight plan for every backyard or commercial drone capable of entering the nation's airspace, then that's what it will mean.

The laws are still unclear, though the government has recently affirmed that the FAA does indeed have the right—and, in fact, the obligation—to regulate unmanned air traffic. At press time, the law in much of mainland Europe is much clearer: You need a certificate to fly *any* kind of drone, *anywhere*. In the United States, the FAA has proposed what appear to be reasonable rules, though they may not meet with favor from vendors such as Amazon, which hopes to use a fleet of drones to deliver goods. The current FAA proposal requires that small commercial or private drones be flown only in line of sight, which would seem, in the United States, at least, to prohibit the sort of semiautonomous drones that Amazon had been envisioning. However, in March of 2015 the FAA did grant Amazon a "certificate of experimental airworthiness" that will at least allow the company to begin testing its drone delivery fleet, so Bezos and crew may yet find a way to fill the skies with delivery drones.

"It's unfortunate that bad decisions made by a few are going to limit the freedom of many," says tech editor and drone enthusiast Chris Angelini. "Laws will be put into place dictating where you can fly them, at what range and height, and maybe even what equipment they can use. It's too bad that common sense won't prevail here; but as with laser pointers [as noted, recently and increasingly used to blind pilots on approach], when enough people abuse the technology, deterrents will have to be used."

DJI's Michael Perry is hopeful that regulatory intrusions will be minimal.

"Certainly any moving object—including cars, planes, boats, helicopters, even bicycles—runs the risk of causing damage. The good news is that the personal and property damage caused by small unmanned systems is comparatively smaller than these other vehicles I mentioned."

Perry continues: "It's a brand-new field, and there is still much that needs to be done. A lot of these risks can be tackled through the intersection of education, improved technological controls (e.g., DJI's "fly safe" zones set around airports), and intelligent regulation. We're seeing progress on all of these fronts worldwide as people are getting more familiar with the technology."

One hopes that the solutions found will not be terribly draconian, but the upshot is that solutions *must* be found.

And, as with all regulatory developments affecting technology, we can hope that it's possible to steer a careful path, helping to ensure safety and privacy while still allowing citizens relatively unfettered access to the technology.

After many months of dithering, the US government finally did come up with rules requiring the registration of hobby drones. In mid-December of 2015, the FAA notified the public (and drone manufacturers) that drones weighing less than fifty-five pounds would need to be registered. (The FAA is charging a $5 registration fee. Drone operators can register their aircraft—note that one registration can cover multiple drones—at faa.gov/uas/registration.) In the future, newly acquired drones must be registered prior to their first flight, in much the same way that certain types of walkie-talkies and other transmitters must be licensed prior to their initial use.

The issue, notes FAA spokesperson Michael Huerta, is accountability.

"Registration gives us the opportunity to educate these new airspace users before they fly so they know the airspace rules, and understand [that] they are accountable to the public for flying responsibly," says Huerta.

The new rules address hobby and recreation drones; rules covering *commercial* use are pending at press time.

How the Playing Field Has Leveled

Consider what aircraft used to cost, relative to one's income—and keep in mind the slightly astounding fact that they're available at all for civilian use. Just as aircraft costs have dropped ("plummeted" just doesn't seem like a good word to use when discussing aircraft), so have the costs of computers. And what is a drone, really, but a sort of roboticized, computerized aircraft?

Like many other now-ubiquitous technologies, including the Internet, drones began as military tools. They were (and still are) used to attack, to defend, and to gather intelligence. But, also like the Internet, drone technology (many hobbyists prefer the term "multi-rotors" as a way of avoiding the military connotations, says Chris Angelini) has moved far beyond its military origins; now drones are used for fun and for research, for lifesaving and for business. And the reality is that we simply have no idea what uses will be made of drones in the future, or with what other new tech-

When Amazon's Jeff Bezos announced plans to create a drone delivery service for its products, many wrote it off as a publicity stunt, but Bezos is dead serious. He says Prime Air will be ready to launch (so to speak) as soon as regulations have been updated to account for this sort of commercial drone traffic. *Image courtesy of Amazon, Inc.*

nologies they'll be combined to form still more new tools or technologies. (What happens when the first AI drone is built? Who will be the first to program a drone to walk the family dog? A smartphone-controlled drone has access to the Internet; what would happen if we allowed people to pay for the pleasure of controlling from Quebec, say, a drone flying over Calabasas, California? Could drones herd cattle or other livestock? Could they do it autonomously? What would a security drone look like? And on and on, with almost literally no end.)

"The applications we're seeing now," says DJI's Michael Perry, "helping farmers intelligently improve yields, supporting conservationists tracking endangered species, increasing the efficiency of search-and-rescue teams—are unequivocally a good thing. Making this technology easier to access and more reliable will undoubtedly open up even more innovation down the line. As more people get their hands on the technology, the more exciting applications open up. By making aerial capabilities cheaper and more accessible, innovators are just now starting to think about the possibilities of what they can do from the sky."

What's Next?

As we've seen, things happen quickly in technology, and that's particularly true when we're speaking of flight-related technology. How quickly? Well, consider that between the Wright Brothers' first flight and the day man landed on the moon, only sixty-six years had elapsed. In that brief period of time, well within a person's lifetime, we went from a few seconds of skimming hesitantly mere inches above a North Carolina beach, to rocketing men and materials to the surface of the moon, some 240,000 miles distant. We landed them on the surface, allowed them to explore, and then recovered the men and the lander and flew them another 240,000 miles back, where they splashed down safely in the Pacific Ocean. (Keep in mind that the Soviet Union's unmanned *Luna 2* lander, laden with scientific instruments, made it to the moon's surface a full ten years before *Apollo 11*'s manned 1969 moon landing. Then again, *Luna 2* didn't "land" so much as *crash* onto the Moon's surface while traveling at a speed of over 7,000 miles per hour.)

Given the ever-increasing speed at which new technology is developed, distributed, and then improved upon or used in ways no one had even considered, we can rest assured that—for better or worse, or most likely both—drones will rapidly become ever more efficient, less expensive, and more ubiquitous.

Part of what makes drones possible, of course, are the many advances that have taken place in aeronautical engineering since the days of Langley and the Wright Brothers, a few of which have been mentioned here.

But the other thing that makes drones work is the computer. Without the computer and the processing power of allied systems and technologies—smartphones, digital receivers, radio controllers, and the like—the closest we could get to today's sophisticated drones would be the "bug" built in 1917, the Kettering aerial torpedo that simply flew off blindly toward the target (or so its minders hoped) until enough time had elapsed, at which point the wings fell off and the bug became a (not very smart) bomb. But computers controlling (and literally guiding) the craft brought to the enterprise a level of sophistication that would not have been possible otherwise.

In the next chapter, we'll take a look at the computers that made all of this possible, and at the Internet that serves as a communicative backbone for so many of our computerized devices.

Chapter Six

Enhancing Your Brain: Computers and the Internet

Electronic aids, particularly domestic computers, will help the inner migration, the opting out of reality. Reality is no longer going to be the stuff out there, but the stuff inside your head. It's going to be commercial and nasty at the same time.

—J. G. Ballard

The Perfect Storm

In the United States, the winter of 1950 will be remembered chiefly for a terrific Thanksgiving storm (some have called it a "perfect storm") that slammed into the US East Coast, leaving death and destruction wherever it touched down. It produced widespread flooding and wind damage, along with record snowfalls and minimum temperatures over just about all of the northeastern part of the country.

By all accounts, it was awful—a furious tempest of wind, cold, and driving rain and snow that pummeled the entire East Coast for days. It began in the Southeast; one writer noted that the temperature in Atlanta, Georgia, dropped from the low 50s to 3 degrees or less overnight. Then it hit New York (where a 94-mile-per-hour wind gust was recorded), New Hampshire (which saw a wind gust measured at 110 miles per hour), and the Northeast before heading out over Ohio and Pennsylvania. (Pittsburgh was hit with thirty inches of snow, bringing the city—and the country's steel industry—to a shuddering halt.)

It was an ugly storm; some three hundred people died, and there were as many as one million power outages. It was, at the time, the worst storm ever recorded over the United States. And no one knew it was coming.

As one publication noted, "The severity and lack of warning of the storm served as motivation for it to become the subject of the first experiments in Numerical Weather Prediction, which ultimately led to the present-day National Centers for Environmental Prediction (NCEP)."

And that's where the computers come in.

Computerized weather forecasting was in its infancy in 1950, but there were some signs that it was about to be revolutionized. Mathematician John von Neumann published a paper on the use of computers to help predict weather, though their efficacy was constrained by limits in storage and processing capabilities.

Nonetheless, computer-driven weather forecasting was beginning to make some serious strides. By 1950, the processing power and storage (and predictive algorithms) had improved to the point that it was taking about twenty-four hours to predict a twenty-four-hour weather pattern. In other words, computer scientists, including von Neumann, Jule Gregory Charney, and Ragnar Fjørtoft, all working at Princeton's Institute for Advanced Studies, were able to use computers to correctly predict the weather—but only as it happened. The calculations were just barely keeping pace with the weather itself. Such "predictions" were thus too late to have served to warn people about specific storms. (Nonetheless, this was an astounding feat. Back in 1922, meteorologist Lewis Fry Richardson had spent a full year creating a fairly accurate twenty-four-hour forecast. He figured that in order to create a forecast in something close to real time, some 64,000 "calculators"—by which he meant *people*, sitting at desks and manually working out the math—would be needed. And even then, the forecast would only be finished *after* the event.)

But the fact that the timing was off didn't matter, really. It worked. The Princeton group's computers could parse the meteorological data, filter out the irrelevant "noise," and correctly predict what was going to happen—no matter that it had *already* happened, or was happening even as the computer (at first, ENIAC, the Electronic Numerical Integrator And Computer) churned out the answers. The rest was just technology: better algorithms, more storage, faster processing.

A few years later, more-advanced computers (somewhere between $175,00 and $300,000 per year was set aside to pay rental fees on an IBM 701 computer for the Princeton group) were routinely forecasting local and regional weather twenty, thirty, and forty days out. At least in theory, the crippling effects and the loss of life occasioned by the Thanksgiving storm of 1950 need never be repeated. (In 1947, RCA's director of research had in fact predicted that computers would not only predict the weather, they would soon *control* it. We're still waiting for that one.)

The Birth of the Technology

But where did they come from, these computers?

When discussing the birth of computers, we often point to John Mauchly and J. Presper Eckert's mid-1940s invention of ENIAC at the University of Pennsylvania, a fifty-ton behemoth that utilized more than 17,000 glowing, flickering vacuum tubes (several of which burned out every day, requiring technicians to spend a great deal of time locating and replacing the finicky devices).

Others might point to the British effort that resulted in Colossus, a machine that was kept secret until the 1970s. Its secrecy was a shame, really, because many of the people who worked on Colossus contributed a great deal to the fledgling discipline now known as computer science, but their contributions were overlooked for many years. Preceding ENIAC by three years, Colossus was really the very first digital, electronic machine that was programmable and could be used as something like a general purpose computer.

Sometimes we look to English philosopher and mathematician Charles Babbage, who in the 1820s designed an "analytic engine," which, had the technology existed to support it, would have functioned as a true programmable mechanical computer. (Some would argue that Ada Lovelace, Babbage's correspondent and protégé—and a formidable mathematician in her own right—was actually the first computer programmer; she wrote instructions or algorithms that would have been run on Babbage's machine, had it been built.) Sometimes we go even further back, perhaps as far as the Jacquard loom, invented in 1801—and which itself owed much to even earlier inventions going back at least to the mid-eighteenth century or before. The loom used what is recognizably a form of punch card to enable the weaver to repeat patterns reliably and just about endlessly.

There's an argument to be made, though, that the birth of the modern computing age owes something to the spectacularly unsung Robert Percival Porter.

Porter was a journalist, a prolific writer, and a part-time economist. (Yes, an economist; the "dismal science," the main function of which is "to make astrology look respectable," as John Kenneth Galbraith and others have said.) Born in England in 1852, Porter immigrated to the United States as a child, eventually becoming a journalist and then a member of President Chester A. Arthur's Tariff Commission, a committee established by Arthur to determine just how much tariffs should be reduced in order to encourage trade. (The resulting legislation was so complicated and so politically unpopular that it became known as the Mongrel Tariff Act.) Porter was also cofounder of the *New York Press* and wrote books on a mind-numbing variety of topics, including a hagiographic biography of President McKinley, as well as dry tomes discussing the commercial and industrial status of Cuba and the rise of Japan as a modern power, most of which were widely ignored.

Among his many other accomplishments, Porter was for several years superintendent of the Census, appointed at a time when the government and its census-taking activities had encountered a serious problem: Such was the rate of growth of the country that the previous superintendent, Francis Amasa Walker (yet another economist-cum-journalist), had taken fully eight years to complete the 1880 census. This was not completely unexpected. As the country and its population expanded, that expansion was accompanied by a deluge of data, and the collection, tabulation, and coordination of this data was becoming alarmingly difficult; simply keeping track of the country's population had in fact become almost impossible. It was apparent that, using existing methods, Porter's 1890 census would not be finished in time to begin work on the 1900 census. (Keep in mind that the census was—and is—quite important: It's mandated by the US Constitution both to apportion taxation between the states and as a way of determining congressional representation. The country desperately needed this information, and the data had to be both timely and accurate.)

Thus, a problem loomed, and Porter and others in the federal government sought a solution. (Their level of desperation was such that, for the first time, the government allowed women—frail and delicate though they were thought to be—to act as "enumerators," or collectors of data.)

Adding women to the workforce certainly helped (after all, it meant that more people were out there collecting data, and the women, naturally, were paid less than the male enumerators), but the real solution turned out to be technological: A former Census Bureau employee named Herman Hollerith had invented a tabulating machine that used punched cards to collect and tabulate census information (or, ultimately, any kind of data) quickly and efficiently.

Hollerith had a pedigree that made him the perfect person to find a technological solution to the Bureau's problem. A mining engineer at a young age, he earned a PhD from Columbia and taught mechanical engineering at MIT. He had for years been experimenting with using punch cards as a data-storage mechanism.

Hollerith's tabulating machine was a brilliant idea. While the Census Bureau's relationship with Hollerith would eventually sour (Hollerith kept

After several acquisitions and mergers, the company started by Herman Hollerith would be run for decades (starting in 1914) by Thomas J. Watson, Sr. In time, the company would become known as IBM. *Image licensed under Creative Commons CC-BY-SA-3.0.*

raising his prices until the Bureau had no choice but to build its own electromechanical tabulators), Porter recognized a momentous technical achievement when he saw one, and he knew how to take advantage of it; he ordered dozens of the machines from Hollerith's newly formed Tabulating Machine Company. The Hollerith tabulators made it possible for one Census Bureau employee to tabulate all the data collected by thousands of enumerators. (About twenty years later, through a series of acquisitions and mergers, Hollerith's company would become International Business Machines, or IBM.)

It was a brilliant move. Thanks to Porter and Hollerith, the Bureau finished the 1890 census months ahead of its deadline and considerably under budget. (In a neat piece of irony, the designers of the Census Bureau's new tabulating machine, the one Hollerith forced them to build by raising his prices until his machines finally became unaffordable, started their own company to compete with Hollerith's; the latter company would eventually become part of the Remington Rand Corporation.)

Hollerith is widely accepted these days as one of the fathers of computing; it was his machine, after all, which showed that permanent data storage—in this case, punched cards—could be used as input for a computing device that could perform complex tabulations thousands of times faster than could be accomplished by any human. But it was the virtually unknown Robert Percival Porter who recognized the import of these devices and who was willing to put his job on the line by recommending that the Census Bureau use the

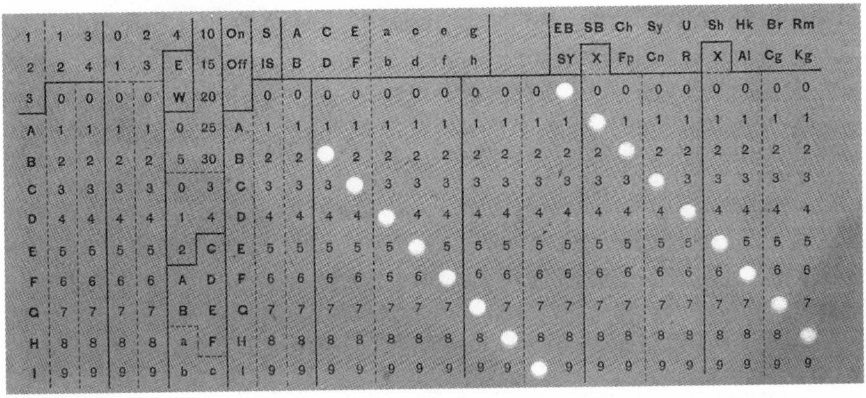

A Hollerith punch card used to tabulate data. *Public domain image courtesy of the Library of Congress.*

unproven machines to complete the 1890 census. Surely this makes Porter, if not one of the fathers of modern computing, then at least one of the godfathers, or perhaps an especially close uncle.

Computers for "The Rest of Us"

And then things got crazy, fast.

We went from tennis court–size, multiton behemoths such as EDVAC, the Electronic Discrete Variable Automatic Computer, and ENIAC to better, faster, and *smaller* computers, very quickly.

In 1948, astronomer Wallace Eckert (no relation to J. Presper Eckert) helped IBM to create IBM's Selective Sequence Electronic Calculator (SSEC), which was used to create moon-position tables that would eventually be used by the *Apollo 11* flight to the moon in 1969. In 1950 (even as scientists at Princeton's Institute for Advanced Studies labored to produce accurate computer-derived weather forecasts), a Minneapolis company produced the ERA 1101, which can lay claim to being the world's first commercially produced computer. (The company's first customer was the US Navy.)

The following year, Remington Rand delivered to the US Census Bureau (back to them again) the UNIVAC I. The company sold forty-six of the machines at the startling cost of about $1 million each. (To be fair, the $1 million price tag did include a high-speed printer.) The year 1954 saw the introduction of what may have been the world's first mass-produced computer, the IBM 650, of which 450 were sold in one year.

All along, improvements had been taking place. Each new device had better, faster memory, faster processing, and more—and more accessible—storage. One key improvement resulted from the 1947 invention at Bell Labs of the transistor. A team consisting of William Shockley, John Bardeen, and Walter Brattain created a "point contact"-type device that was quickly superseded by the much more efficient "bipolar" transistor. The result was that circuits could be greatly reduced in size, and the heat generated by the circuit itself could be similarly reduced. The transistor did the job of a vacuum tube (allowing a weak signal to control the flow of a much stronger one), but it did so much more elegantly and efficiently. Suddenly, electronic devices such as radios—heretofore somewhat bulky affairs—could be made stunningly small. (Baby boomers will recall the advent of the transistor radio, tiny—or so they seemed at the time—plastic boxes from which issued tinny-sounding versions of the era's popular music, and which were held to the ear of just about every teenager one encountered. Teens have *always* embraced technology, and almost always to the consternation of parents and teachers.)

IBM's 1959 7000-series computers were the company's first mass-produced transistorized machines, and the following year, DEC released a similarly transistorized (and thus much smaller and more reliable) computer, the PDP-1, which sold for a relatively affordable $120,000. (It's said that the first computer game, *SpaceWar!*, was written on the PDP-1 by hackers at MIT.)

Things continued along a predictable path for several years: Large companies produced ever smaller and more efficient mainframe (and then, mini) computers, which they sold—at steadily falling prices—to universities, government agencies, and larger corporations. (And transistors, which had revolutionized electronics in the early 1950s, were themselves on the way out, beginning to be replaced in the 1960s by integrated circuits in which the functionality of many transistors was built into a chip, or "package.")

In 1971, John Blankenbaker, an unemployed engineer, released a small computer he called the Kenbak-1. It was simple and relatively inexpensive (about $750), though functionally very limited. It used lights as both input and output, had a 256-byte memory, and was programmed in assembly language, a low-level language specific to the machine's architecture or microprocessor. (The Kenbak-1 did not actually utilize a microprocessor, since none were available at the time.)

Simple though it was, about forty or so Kenbak-1 computers were sold, mostly to schools that used them as teaching devices in courses aimed at beginning computer science or engineering students. According to some, the Kenbak-1 is in effect the first "personal computer." It was followed in 1972 by the Micral, produced by the French company, R2E. The Micral was the first commercial computer based on a microprocessor, in this case, the Intel 8008.

In 1974, researchers at PARC (Xerox's Palo Alto Research Center) designed—but never sold—the Alto, a machine that was destined to revolutionize personal computing, though indirectly. Considerably ahead of its time, the Alto (and the later Xerox Star) featured a mouse, graphical menus, icons, and a windows-based display. (It would be Steve Jobs's 1979 visit to the Palo Alto facility that would set Apple on a track that led the company to develop the Macintosh and Lisa computers a few years later.)

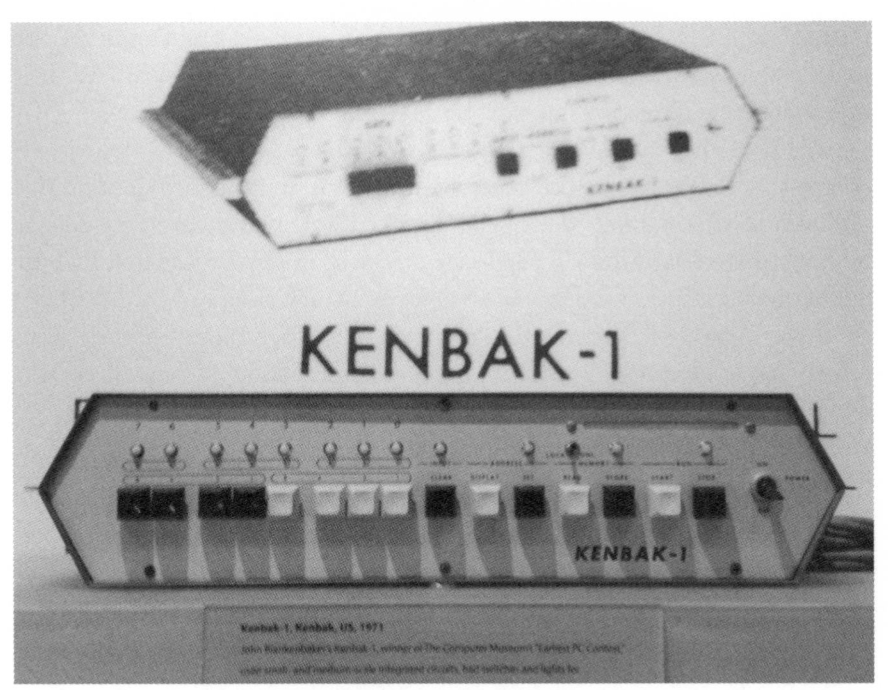

The Kenbak-1 may have been the first personal computer, though it was intended at the time as an educational tool and sold largely to schools. *Image licensed under the Creative Commons Attribution-ShareAlike 2.0 Generic License. See http://creativecommons.org/licenses/by-sa/2.0/deed.en.*

In the mid- to late 1970s, things were beginning to heat up. Early in the decade, engineer (and medical doctor) Ed Roberts formed MITS (Micro Instrumentation and Telemetry Systems), and then, a few years later, created the Altair 8800 personal computer, a kit that found a wide audience with engineers and hobbyists. Bill Gates and Paul Allen wrote Altair BASIC for MITS, thus creating Microsoft's very first product. (At the time, Gates would have been a fresh-faced twenty years old and a recent Harvard dropout.) Only a year later, Steve Wozniak designed and built the Apple I computer, and Woz and Steve Jobs produced about two hundred units before moving on to the Apple II, one year later.

By 1977, the home computer landscape—which had not even existed a few years before—was booming, dominated by the Apple II and the

Commodore PET. (PET stood for Personal Electronic Transactor, though no one seemed to know—or care—about that.) Both the PET and the Apple II utilized cassette-tape drives as storage, though single-sided 5.25-inch floppy drives would be released shortly. The PET came with 4K or 8K of memory, while the Apple came with 4K, but could be expanded to an astounding 48K. In that same year, Tandy (Radio Shack) would jump on the bandwagon, with its TRS-80. The company had expected to sell about 3,000 units, and was completely unprepared when orders for some 10,000 computers came rolling in within a month of the computer's release. Atari joined the party in 1979 with its Model 400 and Model 800 machines, the former of which was marketed mainly as a gaming system.

Your three-pound laptop owes its origins to Adam Osborne's Osborne I "portable" computer, introduced in 1981. The first "luggable," the Osborne I, with its five-inch display, weighed twenty-four pounds and cost almost $1,800, but it *was* portable. More or less.

In 1982, Commodore released the Commodore 64, an affordable color system that featured 64K of memory. The C64 was destined to become the most popular—or at least, the best-selling—computer of its time, with sales of twenty-two million units.

The following year, Apple introduced its first flop: the Lisa. Featuring a graphical user interface (GUI) that owed much to PARC's Alto and Star computers, the Lisa came with 1MB of RAM, a twelve-inch monochrome monitor, two floppy diskette drives, and a 5MB hard drive. It was an incredible machine, but it was slow (it takes a lot of processing power to push those pixels around) and expensive ($10,000), and did not have much marketplace impact.

In 1984, in a personal computer world awash with IBM PC clones (the IBM PC was released in 1986, and it included an operating system the company had licensed from a small software company called Microsoft), Apple released the Macintosh, a smaller and much less expensive version of its failed Lisa. The newer system cost $2,500 and included such applications as MacPaint and MacWrite. Like the Lisa, it utilized a GUI, a mouse, and windows. Positioned as the anti-IBM (which was perceived as very staid, very corporate), it was extremely popular, and versions of it are still being sold, though now mostly in either all-in-one or portable/laptop form.

The next several years saw more releases of ever-more-powerful systems (the Amiga, the IBM PS/2, etc.), but the personal computer landscape

The Macintosh, with its graphical user interface, redefined personal computing for generations of users. This is the Macintosh Plus, introduced in January of 1986. *Image courtesy of Rick Brown.*

would ultimately be transformed by personal (and this time truly portable) computers in the form of smartphones and tablets (see chapter 9).

The Internet

The other transformative computer-related technology was the Internet, created in the 1960s as part of a military network called ARPANET.

Developed by the United States' Advanced Research Projects Agency, the first four nodes (four universities: UCLA, Stanford [actually the Stanford Research Institute], UC Santa Barbara, and the University of Utah) were connected in 1969. The original goal was not so much to provide alternate communicative paths in the event of a war, as many have said, but to share resources among scientists at the various universities that would eventually be connected to the Internet.

In 1972, a demo of the Internet was given at the International Computer Communication Conference (ICCC). E-mail was invented that same year (by a contractor named Ray Tomlinson), and then its usability expanded by MIT researcher Lawrence G. Roberts, who wrote a utility to list, read, file, and forward messages. (Tomlinson died in early 2016.)

The Internet is essentially a network of networks; each network is independent, but connected to one another—and to the Internet at large—through the use of agreed-upon protocols and applications developed by networking pioneers such as Bob Kahn, Vint Cerf, Howard Frank, Leonard Kleinrock, and many others.

With the development of LANs, PCs, and workstations in the 1980s, the Internet was becoming a busy, crowded, and, most of all, *useful* space. Government agencies, scientists, and universities were able to communicate, share information, and collaborate as never before. But the Internet in the 1980s still remained out of reach of most people, and there was certainly no commercial employment of the "network of networks." Most people had not even heard of the Internet.

That changed in 1990, though few were aware of it at first. Tim Berners-Lee (now Sir Tim Berners-Lee), an independent contractor working for CERN (in English, CERN is the European Organization for Nuclear Research; in French, CERN stands for Conseil Européen pour la Recherche Nucléaire) and Belgian computer scientist Robert Cailliau proposed the creation of a hypertext system by which documents could be linked to one another. Berners-Lee built a web browser (the first web browser, in fact), a website (yes, the first website), and tested his creation in December of 1990. The World Wide Web (WWW) had just been invented.

And not just invented, but invented for a specific reason.

"It was designed in order to make it possible to get at documentation," said Berners-Lee in a 2007 interview, "and in order to be able to get

This NeXT workstation was used by Tim Berners-Lee as the first web server on the World Wide Web. It is shown here as displayed in 2005 at Microcosm, the public science museum at CERN. *Image licensed under the Creative Commons Attribution-ShareAlike 3.0 Unported License.*

people—students working with me, contributing to the project, for example—to be able to come in and link in their ideas, so that we wouldn't lose it all if we didn't debrief them before they left. Really, it was designed to be a collaborative workspace for people to design on a system together. That was the exciting thing about it."

From the beginning, then, the web was all about collaboration—the sharing of data, of information. As has been noted elsewhere, this is why the web is not particularly secure or private. It wasn't *meant* to be. We can (and do) attempt to make it more secure, but we're fighting technologies that were never designed with security in mind.

And we have come to rely on it to an extent that we may not really appreciate until we're forced to do without it. Just ask the residents of Flagstaff, Arizona.

On February 25, 2015, a vandal sliced through a fiber-optic cable lying near a riverbed in the desert north of Phoenix. The resulting outage affected cellphone, Internet, and TV service for most of Flagstaff and parts of Phoenix, Prescott, and Sedona. College students at Northern Arizona University were in tears, unable to purchase books, turn in assignments, or view online lectures. In most cases, residents had no Internet access for a full day. In Flagstaff (population 69,000), almost *everything* stopped working: not just TV and Internet, but most cell service (including some 911 calls, which had to be rerouted) and point-of-service systems, including cash registers. Even the ATMs in town were out of service. Why? Because all of it is dependent on the Internet.

The service was back up in fifteen to twenty hours, but what would have happened if the outage had lasted fifteen to twenty days? Or weeks? What if it *never* came back on? We have built an amazing tool, but it's as fragile as it is indispensable. That's a little scary.

THE TOOL THAT DELIVERS . . . WELL, MORE TOOLS

The Internet is a collection of technologies working together to provide a platform for the delivery of content and services. Originally envisioned as an instrument of scientific research, it quickly became much more than that. The Internet (or, more accurately in our case, the World Wide Web) has become ubiquitous and powerful, politicized and commercialized. For better or worse (or both), it is the tool used by teenagers looking to connect with peers, by terrorists looking to find recruits or communicate with cells, by grandmothers in search of genealogical information, by businesses looking to sell products, by journalists seeking information, and much, much more. It has become *the* communicative backbone of our society. (In 2011, the United Nations declared Internet access "a basic human right" and argued that access to it ought not be restricted, even to quell unrest. The UN declaration noted that the writers considered "cutting off users from Internet access, regardless of the justification provided, including on the grounds of violating intellectual property rights law, to be disproportionate and thus a violation of article 19, paragraph 3, of the International Covenant on Civil and Political Rights.")

Given all of that, it's a shame that the web is suspect on several counts.

The first issue is one we've already mentioned: The Internet is porous and unsecured. There is very little about the web that's truly private, and lacking even basic protective measures, it can be used as an offensive weapon to deliver malware, scams, hoaxes, and the like.

Second, since there are essentially no barriers to entry, any idiot can put just about anything up on a website or blog or Facebook page, and if it's done well enough (or if the consumer is naive enough), it can pass for useful information that has the same standing as information delivered by any large news agency. We have democratized information, but in the process, we have also democratized *misinformation.*

That said, the Internet is also a very valuable tool—possibly one of the most valuable we've ever seen. It delivers products, services, and information, and it creates opportunities as it does so. It has democratized access to information and also leveled the playing field to the point where would-be entrepreneurs can go head-to-head with well-funded, established companies; almost anyone can design and create a functional website for very little money, and with some design and coding expertise, the result can look every bit as good—and function just as well—as a site provided by Neiman-Marcus, Best Buy, Nordstrom, or JCPenney.

The Internet is a disruptive force, and its growth has impacted industries as varied as shipping, manufacturing, and publishing, and not always in a good way. But along with that disruption, the Internet has also created jobs, and sometimes entire industries. It may have made some jobs obsolete, but there's no shortage of Internet-fueled entrepreneurial opportunities.

Colorado Crafted: Artisan Gift Boxes

Take Colorado Crafted, founded by two Longmont, Colorado, friends, Dulcie Wilcox and Sarah Welle. They use a website to sell gift boxes filled with local Colorado artisan goods: popcorn, tapenade, soaps and balms, coffees, candies, and more. It's yet one more example of the Internet providing those with an entrepreneurial bent the opportunity to do business by removing many of the barriers to entry.

"I absolutely think we hold our own, design-wise, with the bigger competitors," says Wilcox. Partly that's because the duo uses a set of e-commerce and web-design tools provided by Shopify—which is of course *another* example of a product and service being provided by the web itself. Colorado Crafted's website looks every bit as sophisticated as, say, JCPenney's, mainly

Sarah Welle (left) and Dulcie Wilcox (right) started Colorado Crafted, a web-based business that sells gift boxes containing artisanal goods from the state of Colorado. *Image courtesy of Dulcie Wilcox.*

because another small company, Shopify, created web-deployed tools that Wilcox and Welle used to design and deploy their e-commerce site.

"With Shopify, it has been fabulously easy to have a great-looking, super-functional site, and we have never had to worry about glitches," says Wilcox. "I have a friend who was getting estimates of $4,000 to $8,000 for a website design for his wife's jewelry design business. I told him that we bought our Shopify theme (the basic infrastructure and design for the site) for $140. Next time I spoke with him, he was working on a Shopify website."

Wilcox is talking about a specific vendor she happens to use and with which she seems quite satisfied, but there are many companies that have taken to the web to create and deploy tools that make it easy for the *rest* of us to take to the web. These include such code-free website builders as Jimdo, Weebly, Wix, Squarespace, and others.

The business environment has been democratized by the Internet. Almost anyone can create an e-commerce-enabled website for very little money (or in some cases, for free) and be able to market their products and services around town or around the world.

Learnist: Your Own Personal Researcher

Then there's Learnist, a small web-based "cottage content industry" in which researchers and writers introduce a topic and then provide relevant links to information about that topic. In effect, it's a way to aggregate and curate content. If, for example, you're interested in Buddy Holly, you could sift through all two million Google hits that result when his name is entered (and good luck with that), or you could go to learnist.com and examine professor Iris Mohr's brief overview and check out the links she's already selected and vetted. It's a time-saver and a way to take advantage of someone else's expertise. It's also a service (monetized for the more popular contributors) that didn't exist a few years ago—one more example of a burgeoning Internet subeconomy.

Crowdfunding

One of the most conspicuous examples of technology's democratizing effect can be seen in the various crowdfunding opportunities available via the Internet. Companies such as Kickstarter, GoFundMe, Indiegogo, Fundable, and others provide platforms an individual or a small group can use to raise funds to change a dream into a reality. The dream may have been to create and market the perfect bicycle, an open source computer, a film, a wireless charger, the world's greatest bike rack, or any number of such items. Perhaps the dream was to raise funds for an orphanage. Or for a honeymoon that your friend's aging grandparents never got to take. All of these and more have been funded by people, many of them unknown to the recipient, and many of them contributing only a few dollars. And because the Internet provides a platform for such funding, many potentially life-changing (or even lifesaving) products are seeing the light of day, many artistic endeavors are being produced, and many dreams are being realized.

Crowd Supply is one of those crowdfunding platforms. Begun in 2013 by MIT grads Joshua Lifton and Scott Torborg, along with entrepreneur and investment banker Lou Doctor, Crowd Supply sees itself as occupying a niche that's different than that occupied by Kickstarter and some of the other crowdfunding platforms.

"We started Crowd Supply to be the crowdfunding platform we'd want to use ourselves," says Lifton. "We saw that a huge set of needs wasn't being met in this space, and that products that could have otherwise been quite successful ended up failing because of it."

Lifton points out that Crowd Supply is essentially a tool for product development, while Kickstarter is largely devoted to arts patronage.

"They're both creative endeavors," he says, "but they are fundamentally different in their needs and execution. Crowd Supply is geared toward engineers and designers who create tangible goods that then get sent to their backers. It doesn't matter if our creators are making hot sauce, bicycles, jackets, electronics, or books, as long as the end result is tangible and can be shipped to their customers. All products need initial funding, marketing, fulfillment, and ongoing sales. Crowd Supply provides all those things and more because our creators have demanded them, and because their success is our success."

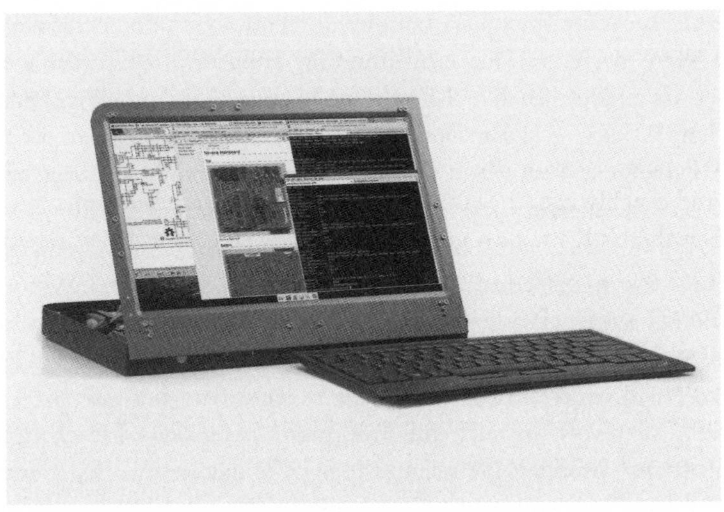

The Novena open-source laptop may be Crowd Supply's most successful campaign. At over $780K, over triple its goal of $250K, it's the company's highest-grossing campaign. *Image courtesy of Crowd Supply.*

Crowdfunding in general, and Crowd Supply in particular, is a good example of the democratization of technology, and of how that democratized technology is then used to further democratize other aspects of life—in this case, crowdsourced business investments resulting in the creation of a product.

Says Lifton, "Crowd Supply helps to democratize product development in many ways, all of which stem from a core tenet of crowdfunding: low-risk market validation. This is good for everyone. It allows creators to make an informed choice about how much time and money they should devote to a product before they actually go through a full production run. It allows backers to give direct feedback to creators so that the resulting product is more likely to suit the backers' needs. It lowers the barrier to entry so that people who wouldn't have otherwise designed and manufactured a product can now do so."

Using Technology to Do Good

And let's not forget that crowdfunding can be used for purposes other than purely commercial. People can get together to do good things—sometimes wonderful things—for other people, for animals, for the world.

Take the EARTH Animal Sanctuary in Roberts, Illinois. Founded in 2013 and home to a large number of rescued animals, the organization "offers sanctuary to all species without discrimination." Including pigs, such as Bentley, a young potbellied piglet who turned out to be in need of some very expensive veterinary care. The organization turned to a crowdfunding platform called YouCaring.com to raise funds to cover Bentley's medical costs, and quickly raised well over $20,000 to pay Bentley's bills.

Then there's Kiva, a nonprofit crowdfunding site that's aimed at micro-finance—finding individuals to lend sums as small as $25 "to help create opportunity around the world." Founded in 2005, at press time Kiva had secured over $693 million in loans in eighty-six different countries—and all with a repayment rate of over 98 percent.

Robert Keith Froom runs a nonprofit called 4 the World. Its mission is to improve education and health (two things that are in fact connected in various ways) in developing countries. The Internet has turned out to be one of the organization's most valuable tools.

"I started 4 the World in 2004 after a trip to Belize where I'd asked a group of kids, 'If I were able to help you in some way, what is it that you would need or want me to do?' The children asked me to help their school so that they could get a better education that would lead to employment and a chance to escape a life of poverty."

Froom returned home to the United States and started 4 the World. After reaching out to family, friends, and local businesses for support, he was able to return to Belize and refurbish a school and build a feeding center. After that, Froom was hooked; he dedicated his life to bringing education and health care to poverty-stricken parts of the world like Belize and Guatemala.

"In Guatemala, I have told students that man has been on the moon and the students did not believe me, so I pulled up a YouTube video and started showing them an astronaut walking on the moon, a space shuttle launch, and even a female astronaut piloting a spaceship. I turned around and was struck

Robert Keith Froom (pictured above) started 4 the World to bring education and health care to third-world countries. Because many Central American schools are so overcrowded, this Guatemalan school is a primary school in the morning and a high school in the afternoon. This classroom was added to provide more room, although the students have to use old, broken desks. *Image courtesy of Robert Keith Froom, 4 the World.*

by the young Guatemalan children staring at me in amazement. Through the Internet we could even connect students from the USA to a classroom in Guatemala, and also connect the kids in Guatemala to their new US friends who were each trying to learn the other's language. All of these things were possible only through the Internet, and it changed my understanding of the importance of [the donation of] even a single used laptop, and how it could change a group of children's lives forever."

Froom believes that all the world's problems can be solved through education, and that computers and the Internet are a big part of finding those solutions.

The result is clear: We've used technology to lower the barriers to entry so that people who heretofore could not have done something (in this case, design and manufacture a product, start a business, save a sick animal) can now do so. And that is the very definition of a more level playing field.

Knowledge and Information

Given the online availability of information about almost everything, it's been said that ignorance is now a choice. That is, if there's some fact you do not know, some answer you do not have, it's largely your fault; the answer is out there—you've just not yet chosen to look for it.

And there's some truth to that. It's difficult to come up with a fact-based question to which you could not come up with an answer based on Internet sources. Will it be the *correct* answer? Well, not always. It pays to become an intelligent web consumer; one quickly learns that not all Internet sources are accurate, objective, or trustworthy. But telling the useful answers from the garbage can be difficult. After all, we've removed all of those barriers to entry; just about *anyone* can put anything he wants up on the web. On the web we can find out that John F. Kennedy was the youngest president, that the Great Wall of China is the only man-made object visible from space, and that President Obama was not born in the United States. Of course, none of those statements is true, but they're out there on the web, often masquerading as fact.

The Internet is as much an information minefield as an information gold mine.

Finding truth on the Internet in not always easy or simple. Many teachers dealing with this subject encourage their students to use what's called the CRAAP system. It consists of asking five questions: 1) Is the information *current*, or is the site old and disused, littered with dead links? 2) Is the information *relevant* and aimed at the appropriate audience? 3) What is the *authority* for the information? Who is the author? The publisher? Are they qualified? Is there contact information provided? 4) Is it *accurate*? Where does the information come from? Is it supported by evidence? Has the information been reviewed or refereed? 5) What is the *purpose* of the information? Is it fact? Opinion? Propaganda? Are there biases?

By now, we've become used to people being fooled by "craap" they've read on the Internet. Even journalists, desperate for a scoop and often under deadline pressure, have been duped by misinformation they've seen on the web. These are people we count on for accurate, unbiased information; of all people, journalists need to be careful about verifying "facts" they encounter on the web. (This is why the European Journalism Centre has published specifically for journalists a *Verification Handbook* that explains how to ensure that information encountered on the web, especially in the wake of some disaster or emergency, is truly accurate and unbiased.)

And now a word about Wikipedia: *Get over it.* All right, so *three* words. Many of us (especially those of us who teach) tend to have a knee-jerk reaction (normally a negative one) to the idea of people in general—and students in particular—using Wikipedia as a source. They—we—need to get a grip. It's just data. In this case, crowdsourced data. It's not always accurate, but usually it can be relied upon. That's because Wikipedia is basically self-healing: People do put misinformation up on Wikipedia, but it usually gets discovered and corrected very quickly. (Though not always. There have been reports of hoax articles staying up for weeks, months, or even years.)

In the end, Wikipedia may be the world's greatest experiment in the democratization of information. It truly is an attempt to provide—essentially for free, though the organization occasionally asks for donations—access to a literally endless supply of information, all of which has been vetted by editors, many of whom happen to be subject-matter experts.

"Wikipedia was founded on the idea that the sum of all knowledge should be freely available to everyone," says a spokesperson for the Wikimedia Foundation, "and that anyone can contribute to that knowledge. The open, collaborative model was designed to allow anyone to create, curate, and access knowledge without restriction."

More than 75,000 contributors from around the world edit Wikipedia each month, and, as of 2013, there were some 4.3 million articles in the English-language version. The editors and contributors come from all walks of life and represent a diverse mix of cultural, socioeconomic, and political backgrounds, and with each edit, they make Wikipedia more accurate and all-inclusive.

It's a truly impressive and largely accurate informational resource, and very popular: In August of 2013, it served 496 million unique visitors.

Now, all of that said, I wouldn't ever rely on Wikipedia as the *only* source of a particular fact. Then again, journalists (and students and other researchers) should always double- (or triple-) source facts in any case. But Wikipedia has proven to be an excellent place to get some basic background and to *start* looking for more in-depth information. And if nothing else, the articles themselves are footnoted and sourced; students (and the rest of us) can use the Wikipedia article as a starting point, and then examine the article's sources. (I have actually encountered a couple of published authors who cited Wikipedia as a primary source; I'm afraid that I'm not yet capable of being quite *that* open-minded, but we're apparently headed in that direction.)

On the downside . . .

The Internet: Democratizing Bullying and Harassment

Actress Emma Watson has emerged as an eloquent spokesperson for human rights and gender equality. Since her *Harry Potter* days, Watson has addressed the UN, become UNICEF's UN Goodwill Ambassador, and has conducted live gender equality–oriented Q&As on Facebook.

Along the way, though, she has somehow managed to raise the ire of Internet troglodytes, some of whom went so far as to send rape and death threats after she delivered a gender-equality speech at the UN.

Sadly, this is not an isolated incident.

Another actress, Ashley Judd, found herself the target of hateful comments after she tweeted that she thought

Actress Emma Watson has emerged as an articulate spokesperson for women's rights and gender equity. She has also been the subject of online harassment as a result of her activism.

members of the Arkansas basketball team were cheating during an SEC conference championship game. Again, this wasn't a situation in which commenters argued with her or accused her of bias or simply pointed out that she was *wrong*. Instead, the "discussion" immediately escalated to what Judd called a "tsunami of gender-based violence and misogyny." (Comments included references to sexual organs, and tweets called Judd a c*nt, a b*tch, and a whore. These are not terms one uses in argument or discussion. They are words that schoolyard bullies employ when on the attack.) As *USA Today* writer

Rem Rieder points out, "There are many wonderful things about the digital realm. But there's a substantial dark side, too. And one of its deeply troubling aspects is the amount of really ugly abuse that some people, from the safety of their computer screens, feel free to direct at those with whom they disagree—or simply don't like. And frequently the targets are women."

There seems to be no end to the anonymity-fueled nastiness of online misogynists. After British activist and journalist Caroline Criado-Perez successfully campaigned to have the Bank of England put a woman on the £10 bill, she was flooded with rape and death threats on Twitter. (Few of these folks know how to be subtle. For instance: "I will find you, and you don't want to know what I will do when I do. You're pathetic. Kill yourself. Before I do." About the best that can be said of this person is that he apparently knows the difference between a contraction and a possessive pronoun.)

These people are not arguing Watson's (or the other women's) points. In fact, they're not *arguing* at all. They're making threats, often physical, virulent, and hateful. (And very often, illegal. Many states have criminal-threatening statutes. Others have antiterrorism laws that can be—and have been—used to prosecute those who exhibit threatening behavior.) They're not saying, "You're very, very wrong about that! Here's where you're wrong . . ." Nor are they saying, somewhat less diplomatically, "Hey, you're just plain stupid! If you believe that crap, then you're a damned idiot!"

Even the latter is . . . well, something *akin* to an argument, at any rate.

But these are not argument, they are *attacks*. They're not aimed at convincing or explaining to you (or to the world) why you're wrong about whatever the topic might be. These are vile and nasty threats, period. They have no place in civilized discourse, but they are encouraged by a medium in which everyone is free to say whatever one wishes with no fear of consequence because it's easy to hide behind a screen name.

And, oddly, such incidents do *both* sides a disservice. The very virulence of the attackers drowns out what may have originally been some very real and valid points of contention, points deserving of debate, of considered argument, of reasonably varying interpretation.

Sadly, this sort of harassment is becoming increasingly common. According to a 2014 Pew report, 40 percent of Internet users have experienced some form of harassment, with users ages eighteen to twenty-nine reporting almost twice as much harassment as users on the whole. The younger users are significantly more likely to say they've experienced physical threats (24 percent), sexual harassment (19 percent), and harassment over a sustained period of time (17 percent). When we consider only young women, the numbers jump even more. Interestingly, online dating sites and apps were *not* particularly likely to be the scene of such harassment; only 6 percent of those who had been harassed said that the harassment had taken place there, while 66 percent said the harassment had occurred on a social media site, such as Facebook or Twitter.

We don't have to go very far to unearth these sorts of attacks; we certainly don't have to look for a rich and famous actress. Anyone can become a victim.

Take Zoe Quinn and Alex Lifschitz, two of the most public faces of the Gamergate controversy.

Gamergate started out being about what some people felt was shady journalism: game reviewers who may have been influenced one way or another (but not by the quality of the game itself) to give positive reviews. At the beginning (for about an hour, maybe) it was an argument about ethics in journalism—surely a subject about which a group of reasonable people could have an intelligent discussion, or even a civilized argument. Then the issue got hijacked by a bunch of needy, abusive, misogynistic children. Instead of intelligent adults arguing their points of view, it became something else. Something ugly and malicious and disturbing.

Alex Lifschitz and Zoe Quinn were the primary (but not the only) targets of the Gamergate hatred that erupted in the summer of 2014. Since then, they've been harassed, threatened, and hacked, more or less constantly. (In response, they've created Crash Override, an online resource for people who are being harassed.)

"We talk a lot about how important the Internet is," says Lifschitz, "but people don't tend to report on how *damaging* people can be, how people exploit those avenues of the accompanying cultural apathy with the actual consequences that the Internet can bring to people."

It's not just a "virtual" problem, says Alex. People can get physically hurt.

"Something like 'swatting' is an extreme case, but very one-to-one. A lot of the ways that these Internet mobs operate is by trying to hurt you in real life. It's trying to get you to shut up and . . . walk away, as if you're going to come back to anything less than utter tatters. It depends on who you are, and what's going on with your life, how they're trying to assault you. But even if it comes to disinformation campaigns, the Internet is where people look for job references, the Internet is where people look to establish credibility with people they haven't worked with before, and if you have a group of people that's dead-set on ruining you, the Internet is really the most efficient place to do it."

Swatting, if you've not encountered the term, is making a bogus emergency call reporting that something dangerous (perhaps a hostage situation or a shooting) is occurring at your victim's home, in the hopes that law enforcement will send a SWAT team there. The (usually) unspoken goal is that the combination of nervous, heavily armed police and frightened victims will lead to injury or death. It's not a prank. As Zoe Quinn notes, it's something closer to "murder by proxy."

The idea that someone—someone real, an actual flesh-and-blood person in the real world—could die should frighten people and dissuade them from such actions, but it doesn't.

"I don't think there's a ton of people that participate in inherently destructive behavior that really think of long-term consequences, generally, for anything," says Quinn. "It's the same mechanics as any mob, but this one happens to be online."

The solution does not lie in restricting—or doing away with—the Internet, says Quinn. The answer, she

Game developer Zoe Quinn became a target of online harassment in the summer of 2014. The Internet campaign eventually became known as Gamergate. *Image courtesy of Zoe Quinn.*

feels, is to require service providers to step up to do their part.

"The problem isn't the Internet—it's poor moderation of it," she says. "The problem isn't anonymity, it's services not actually enforcing their terms of service. All of these problems can be better managed with actual moderation instead of pretending that the Internet is some alternate dimension where your actions don't matter. It's as if the tech has advanced around twenty years, but our conversations around it have only advanced about ten. It's about realizing the ways in which the Internet can be used to hurt people, internalizing that as something that matters, and realizing that the solution isn't to just simply stop using it, but that services need to step up and start actually enforcing moderation."

There is much to argue about online, probably because there's much to argue about in the real world, and the Internet is, after all, merely a reflection of that world. But there is a great deal of difference between an argument and an attack. On the web, it is *safe* to attack; thus, the line between the two blurs very, very quickly. One can attack online with seeming impunity, so bullies find a safe haven on the web. They can insult, attack, and threaten harm—sometimes real, physical harm—and no one does a thing.

In removing barriers to communication, we have enabled the free exchange of ideas. But we have also allowed whiny, petulant, occasionally dangerous children (on whatever side of whatever argument) free rein.

The Internet is broken. But then, most things are broken—at least, a little.

The web is imperfect. And why wouldn't it be? It is, after all, the nature of software systems to be flawed.

The basic issue is simple: The Internet is both complex and inherently insecure. It was designed that way, built that way. It is an excellent tool for *sharing* data, and not very good at all when it comes to *protecting* data. Thus, companies scramble to patch security holes as best (and as quickly and as often) as they can; in the end, though, they're slapping Band-Aids on a flawed infrastructure and, as often as not, the security patches they build to correct a problem cause new problems down the line. (For example, when Mozilla released version 37.0 of its Firefox browser in late March of 2015, it almost immediately had to release version 37.0.1 because the previous release, meant to resolve a security issue, caused a security problem of its own. And thus the dance continues.)

This is not because the companies don't know what they're doing; it's because, as we've seen, the Internet is not a secure platform, and because software is so complex that it's almost impossible to guarantee that it will always behave as designed or to ensure that the designers can take into account every possible variable and every conceivable environmental variation.

On the Internet, no one knows you're a dog. A BAD dog.

Industry analyst Rob Enderle—principal analyst for the Enderle Group, and previously, senior research fellow for Forrester Research and GiGa Information Group—has an interesting, if somewhat radical, take on the problems associated with online bullying, harassment, and misinformation.

"I think to get access to the Internet in a mature culture, it should require a certain amount of mandatory education so that you understand the damage you could do, you understand the risks that you're taking, and you're up-to-date on how you can protect yourself and your loved ones," says Enderle. "And if you're not willing to do that, I think your access should be restricted so you don't harm yourself or others. Much like you would restrict access to heavy equipment if somebody didn't want to train on how to use it properly, for their own safety and for the safety of those around them. I think we should be treating this as a powerful tool, but also a very dangerous one. There are core skills we should all be acquiring that as far as I can tell haven't even been highlighted as a recommended list of things we should learn. We just hand somebody the keys to the Internet and say, 'Go forth!'"

Enderle doesn't explicitly call for a licensing system for Internet access, but his appeal for "mandatory education" would seem to imply something like that. And although retroactively licensing a technology (especially one so widespread) may be difficult, it's not impossible. As Enderle points out, there were cars and roadways (and traffic accidents) long before there were licensing requirements.

Echoing Gamergate targets Zoe Quinn and Alex Lifschitz, Enderle points out that the Internet is *dangerous*.

"The Internet is a weapon," says Enderle. "And as a result, folks need to treat it with some care. You can get harmed on the Internet, and you can do a lot of harm to others; you can destroy careers and lives on the Internet. And I don't think that we as a culture have thought enough about that to really mitigate the problem. We're so focused in many cases on gun control and the rest; I actually think the Internet is far more dangerous potentially than anything we can talk about with handguns or rifles or anything else, because

Rob Enderle, principal analyst for The Enderle Group, likens the Internet to a dangerous weapon, one that really should require training before people—especially young people—are allowed to use it. *Image courtesy of Rob Enderle.*

it can motivate people to *use* those things against others. And it certainly can be used very effectively to destroy lives, and yet we don't really treat it like we would something that is dangerous."

In effect, the Internet is an accelerator, but one that goes in two directions. It accelerates the good—providing for the acquisition and dissemination of information, enabling communication, offering real-time news, creating vibrant economies. And yet, it also accelerates the bad—spreading hatred, falsehoods, and misinformation, and empowering bullies.

But people like Enderle and Zoe Quinn and Alex Lifschitz note that our problem with the Internet is not a technology problem, it's a *people* problem. The technology was created and is run by people, and people can therefore address the problems caused by it. That's what Zoe Quinn was alluding to when she said that the real fault with online bullying lies not so much with flawed technology, but with service providers who are not living up to the terms of their own service agreements. And it's essentially what industry analyst Rob Enderle is talking about when he calls for mandatory training: Like almost any tool, from a hammer to a shotgun, the Internet is potentially dangerous. It needs to be treated with respect; we need to put in place certain safeguards; and we need to require from users a certain basic understanding of the technology, both its creative uses and its hidden dangers.

What Happens Now?

No one knows where the Internet is going. We didn't really know where it was going when we invented it. (We may have *thought* we knew, but we were pretty clueless about its long-term impact, about where it would lead. This tends to be true of *all* technologies: We set them free, and they go places and accomplish things—for good and for bad—that we would have never imagined.)

We have a pretty good idea that the Next Big (Internet) Thing will be the Internet of Things (see chapter 11 for more information about the IoT), but we're not quite sure how—or even *if*—that will work. We're *definitely* not sure where it will lead. Do I really want my refrigerator talking to my stove? What if they get together and order a new dishwasher as a playmate? Do my lightbulbs really need to communicate with a central server at GE? Do I want my car collecting data and sharing it with whoever's willing to pay for it—or clever enough to steal it?

And then what? Maybe robotics? Up to this point, robotics has been an incredibly productive, exciting, and promising field. It's certainly an area that's set to explode, fueled by advances in sensor design, artificial intelligence, processing power, and storage. Our robots are already connected to various networks; it's not at all difficult to conceive of smarter robots connected to even more networks, to more devices, and to one another. Where will that lead? It sounds a little paranoid to say it, but if we build something smarter than we can control, does that bode well for the species? (Sure, it's been a sci-fi staple since *Frankenstein*, but lots of very intelligent people—visionaries such as Steve Wozniak, Elon Musk, Bill Gates, and Stephen Hawking—are beginning to worry about this scenario. We're getting close enough, goes their thinking—and we may be just stupid enough—to get ourselves into trouble.)

Perhaps the thing to stress is that *we* need to be in control, we need to be in charge of our technologies. When they behave badly, *we* need to step in and change things. If that means licensing, well, that's one approach. If it means forcing service providers to honor their own service agreements, that's another. If it means teaching our children that, like the band saw in the garage, the Internet is a handy but potentially dangerous tool, that's yet another.

It's likely that the authors of the Internet Society's "Brief History of the Internet," visionaries and Internet pioneers every one of them, were on target when they considered the future of the Internet:

> *The most pressing question for the future of the Internet is not how the technology will change, but how the process of change and evolution itself will be managed. . . . The architecture of the Internet has always been driven by a core group of designers, but the form of that group has changed as the number of interested parties has grown. With the success of the Internet has come a proliferation of stakeholders—stakeholders now with an economic as well as an intellectual investment in the network.*

We are those stakeholders. We run our businesses on the web. We teach our children using the Internet. We find knowledge, relaxation, entertainment, and long-lost family and friends on the Internet. In the end, our lives hinge upon the interaction of silicon, solder, and circuits. The question yet to be answered is whether we remain in control of those, or whether we end up being controlled *by* them.

Computers are no longer restricted to desktop or laptop models, of course; they're embedded in almost all of our devices and appliances, from our microscopes to our microwaves. (And many of them connect, one way or another, to the Internet.) In fact, it's difficult these days to think of an electronic device that does *not* contain some sort of computer. There are a number of them in your automobile, certainly; one of those may be in a small display on your dashboard that you use to help find your way: Most of us have some form of GPS device that we use in our travels. The story behind that device is a long and convoluted one, and it is to that story we turn next.

Chapter Seven

Finding Your Way: GPS

The future is here. It's just not evenly distributed.
—William Gibson

Getting There

For ages, people have used technology to find their way from one place to another and to help them survive the sometimes perilous journey. From Paleolithic man's leather toolkit pouch (which contained, among other items, rocks used as hammers and stone cores from which to fashion flaked edges and arrowheads) to stargazing seamen and their sextants to today's astronauts living aboard the largely computerized International Space Station, we've always attempted to equip our adventurers with the most recent tools and with devices designed to help them in their adventures.

Of course, it hasn't always worked out. Take Lewis and Clark, for example. Everyone knows this duo as the co-captains (technically Clark was a lieutenant, but Lewis was happy to share command with his old friend) of an expedition sent by President Jefferson to explore the West and to find a water route across North America. The two men spent a fair amount of money on a fascinating and diverse assortment of equipment and supplies, including lead, tobacco, whiskey (of course), fishing tackle, three bushels of salt, and a frighteningly large collection of very harsh purgatives. (Lewis in particular believed in the use of purgatives to cure just about any ailment. One imagines that his men quickly learned not to complain about any ailment less severe than a broken leg. And even then . . .)

But among the items purchased for the trip were several examples of then-new technologies. One was an air gun, possibly made by Isaiah Lukens of Philadelphia, Pennsylvania. Fully charged, the rifle held air pressurized at over seven hundred pounds per square inch and was capable of firing a .31 caliber bullet with roughly the same force as a Kentucky rifle. It impressed the heck out of the Indians, because it was almost silent, did not belch smoke, and needed no powder. To the Indians, it must have seemed miraculous. And as Arthur C. Clarke has noted, "Any sufficiently advanced technology is indistinguishable from magic."

The rifle actually worked well enough, except for one thing: It was prone to firing unexpectedly. Thus, a young woman standing nearby during

Charles Wilson Peale's 1807 portrait of Meriweather Lewis.
Image in the public domain.

a pre-departure demonstration was accidentally shot in the head. Lewis calmly pronounced her wound "by no means mortal" (and no doubt prescribed a purgative).

Another interesting piece of new technology that Lewis bought was a clock made by a Philadelphia watchmaker named Thomas Parker. At $250, the clock was by far the most expensive single item purchased for the trip: In 1803, the cost of this one clock was equivalent to a typical worker's yearly wage several times over.

The clock was important because the group's mission depended on the men (and eventually one young Shoshone woman, Sacagawea, who acted as guide and interpreter) being able to determine their exact location so that they could produce accurate maps of the country across which they journeyed. As any sailor knows, it's difficult to determine one's longitude without an accurate chronometer, so the new Parker clock, guaranteed to be accurate, reliable, and able to withstand the rigors of travel, was crucial to the mission. (We'll get back to the longitude issue momentarily.)

Naturally, the clock didn't work.

In addition, Lewis and Thomas Jefferson spent months designing an iron-framed boat to be covered with animal skins when navigating rivers. With the skins removed, the vessel was meant to be easy to carry during a portage. The boat (to which the men serving under Lewis and Clark referred derisively as "The Experiment") was to be the salvation of the travelers.

The boat didn't work, either.

Somewhat more recently, the crew of the USS *Jeannette* set off in 1879 from San Francisco, California, to what the expedition's backers were inexplicably sure was a warm, lush island at the top of the world in polar waters somehow magically free of ice. With Lieutenant Commander George W. DeLong in command, the expedition took with them the latest in the day's technologies, including the recently invented telephone and several arc lamps, neither of which worked for long. (Although much was learned by the expedition, and during the subsequent searches for survivors, only a few of the men survived the ill-fated attempt.)

On the downside . . .

Loss of Privacy and Dangerous Distractions

While seemingly a benign tool, GPS is in reality no different than any other technology: Like every device from a hammer to a stick of dynamite, it can be misused or abused.

To begin with, *any* item that can be used while driving can be a distraction, and that means everything from a comb to a telephone. (Not to mention the people we encounter who apparently see no problem with applying makeup, shaving, and possibly writing research papers while zipping down the highway encased in two or more tons of steel and plastic.) A GPS (which could in fact be running on a phone, thereby potentially compounding the issue) is actually *meant* to be used while in a vehicle, and there have been thousands of accidents caused by drivers distracted by their navigational devices, cell phones, and stereos. This is why the makers of GPS units and apps are careful to include a disclaimer pointing out that the driver should not interact with the device while the vehicle is in motion. (In the UK, research shows that about 300,000 motorists have crashed while fiddling with a GPS device, while around 1.5 million "have suddenly veered dangerously or illegally in busy traffic while following its directions.")

The more subtle—but perhaps ultimately, more significant—issue with GPS is the privacy issue that it raises.

One example: Police are now utilizing GPS "darts" that attach to vehicles and which are intended to allow officers to track a fleeing vehicle while avoiding many of the risks that can accompany a chase through city streets. This definitely makes our streets safer, but some have pointed out that a dishonest officer could abuse that ability, perhaps tracking a vehicle onto private property or acquiring private information about an individual, all without a warrant.

> Other potential privacy abuses include the misuse of devices meant to track children and pets. While these can help to ensure the safety of your loved ones (including your small, furry loved ones), in the wrong hands, access to this information could be dangerous. What would happen if a predator or pedophile had access to such information?
>
> The privacy issue is a big one, and we've only scratched the surface here. Because geolocation data and other tools can be used together—for good and for bad—we'll deal with more GPS-related privacy concerns in chapters 10 and 12.

In fact, as most of us know by now, the first release of *any* tool rarely works well or reliably, which is why so many of us live by the wise words of an anonymous engineer: "Never buy the first version of anything."

We've come a long way since the days of Lewis and Clark and DeLong. These days, the primary travel aid on which most of us rely is some sort of GPS (Global Positioning System) device—either a freestanding unit mounted on or in a vehicle's dashboard, or a smartphone application that makes use of the phone's integrated GPS chip. How that tool works and the uses we make of it are a fascinating story.

IT'S NOT LIKE WE THOUGHT

But to begin that story, let's first talk about farming.

If you're like many of us, you probably have an idealized notion of farming, fostered in part by exposure to Rockwellian images of apple-cheeked farm wives serving heaping platters of fresh food to equally apple-cheeked husbands and lean (but happy), work-worn farm hands. In this romanticized vision, green and golden fields of corn and wheat are visible through a lace-framed dining-room window, and a far-off tractor makes its unhurried way toward a picturesque red barn in the distance.

Sounds nice, doesn't it?

To go along with this vision, we imagine that farmers, though we know that they work hard, nonetheless live a simple life, a *good* life: Up early and working late, their days are calm and simple. Theirs is a pastoral existence, bucolic and—for all their difficult toil—a peaceful, uncomplicated one. A rudimentary tractor (we picture a 1950s vintage, in Ford gray, with red trim) and a beat-up pickup truck is about all they need to coax from the land nature's bounty.

We've been misled. Or at the very least, we're very much out-of-date.

Dave Grady, farmer, Lathrop, Missouri

After a spending time in college and on the Colorado ski slopes, Dave Grady came back home to Missouri in 1974 to farm. After forty years, he's now cut back, farming "only" some 2,500 acres, mostly devoted to corn and soybeans. Things have changed since he purchased his first simple tractor.

Today's tractors and combines, for instance, are full of computers and sensors, all generating data that can be used either by other equipment or by the farmer himself as he plans how to get the most out of his land the following year. One example is what farmers call "grid sampling."

"These days, all my fields are grid-sampled," says Grady. "There are six samples taken in every two acres. Let's say it's a one-hundred-acre field, so you're taking fifty different areas, two-acre spots, each getting sampled six

Dave Grady at home on his farm. *Image courtesy of Lesley Jackson Scher.*

times. And when you grid-sample that whole field, it gives you an overlay of exactly what's in that field, in this area and this area and this area, rather than just describing the *whole* field, like we used to do. You'd go in and take a half-dozen samples of one hundred acres, and that was your soil sample. Well, now you take a half-dozen samples every two acres."

But sampling data is only part of it; now the farmer needs to *make use* of that data. And the way that's accomplished is by using GPS and specialized software to monitor what's being done where, saving that information, and then using it next year by creating an overlay that tells the farmer exactly what was sown in that particular area, how much seed was used, how much fertilizer, and what the yield was. With that data, the farmer can plan how to treat that specific piece of land during the harvest and during next year's planting and fertilizing.

Says Grady: "You take that data and see how your yields are and then you take *that* information and give it to the fertilizer guy. He sticks that in *his* computer, and when he's driving across the field, he's spreading fertilizer out in different amounts, whatever it's calling for right underneath the truck at that time. If you go over the hill and the soil's a lot better? Well, it doesn't put out quite so much. If you're in a place where the soil isn't as good, it'll put out more."

What Grady is describing is called "variable rate fertilization" and "variable rate seeding." The technique, now widely adopted even by smaller-scale farmers, varies the amount of seed or fertilizer spread in very specific areas depending on last year's results.

Grady is far from the only farmer who's begun using GPS-related technology.

Shane Peed, Lone Creek Cattle Company, Lincoln, Nebraska

Shane Peed is sitting in a modern office building in Lincoln, Nebraska, wearing a tailored suit and looking every inch the successful young businessman that he is. But with his suit he's wearing scuffed cowboy boots, and he keeps looking outside at the natural grasses and native plants that were used to landscape his offices. One gets the impression that he'd much rather be wearing jeans and a baseball cap, out working the fields.

Peed runs Lone Creek Cattle Company, which produces beef for Certified Piedmontese brand, a breed of cattle that in spite of being very lean none-

This GPS-based guidance system can be retrofitted to many different types of farm machinery. It features a built-in GPS receiver and allows the operator to program a path and select one of several guidance modes to steer across the field. It can also be upgraded to provide auto-steering capabilities. *Image courtesy of TeeJet Technologies.*

theless produces tender and tasty beef. And as much as Shane likes working the cattle ranches themselves, he seems to enjoy the farming side just as much. When he talks about using GPS to guide the firm's tractors and combines, he begins speaking animatedly; he loves this stuff, you can tell.

"The biggest thing is that GPS really improves our efficiency," he says. "There are different types of GPS systems. Some of them are completely satellite-based, and we run those; the satellite-based technology we use will give us something like four inches of accuracy. That's about as good as any human driver's gonna do, so we use that for tillage, for discing, and stuff like that."

But even today's GPS, as advanced as it is, can be improved upon. These days, Peed and many other farmers utilize an RTK (real time kinetic) network that uses a satellite GPS signal paired with other radio-frequency information. The result is a signal that allows a tractor, planter, or combine to navigate with an accuracy of one inch.

"We farm on thirty-inch rows," says Peed, "so when we have plus or minus one inch of accuracy, that's *way* better than any human operator would do. So basically, you just set it, and we sit in the tractor and turn it around at the end of the row. Because everything is so precise, there isn't any overlap of rows."

Before the grid-sampling used by Peed and Grady, a farmer might have sampled each quarter of a 640-acre section. So, dividing the field into quadrants, that means sampling each quadrant four times and using that limited data to create a "prescription" for the entire field.

Things are much more granular now, with more precise measurements being taken much more frequently.

Notes Peed: "The way we do it now actually breaks it down into two-and-a-half-acre segments so that we can vary the rate of our fertilizer, using more or less, depending on the fertility of the soil and on what was produced there last year. Then we vary the rate of the seed population so we put more seed in areas that are more fertile."

When farmers look back on previous years' yields and field prescriptions, what they're really doing is using visual overlays combined with a database that keeps track of inputs and yields, all of it tied together with GPS. It's the GPS that keeps track of locations so that the specialized software can note what was planted where, how much seed and fertilizer was used in a particular sample, and what the ultimate yield was.

In other words, farming has come a long way from that Rockwellian vision described earlier. These days, agricultural equipment is festooned with sensors, cameras, and computers used to help data-savvy farmers maximize their yields while minimizing their costs—that is, helping them increase their ROI, or return on investment, as good businessmen have done forever—and GPS is playing a huge part in that effort.

These Are Not the Directions You Are Looking For

As with all technologies, GPS is something of a double-edged sword. It can be used for purposes of which we wholeheartedly approve, and for some things of which we might not approve at all.

In some cases, the results of the misuse of GPS are unintentionally funny. Police officers around the world must appreciate the ease with which GPS seems to land a great many none-too-bright criminals in hot water.

Witness the recent spate of laughably easy recoveries of stolen items occasioned by thieves stealing either vehicles that contain GPS-enabled devices (left-behind smartphones, say, or purpose-built vehicle trackers) or stealing the GPS-enabled devices themselves.

For instance, the proprietor of a Florida moving company recently came to work and found one of his GPS-equipped big rigs missing. The owner simply dialed in the truck's location, called the police, and headed out to a parking lot across town where a group of men were busy stripping his truck. When the police showed up a moment later, the thieves were easy pickings.

In Kenya, some eight thousand miles from Florida, a businessman was relieved of his smartphone and laptop by a pair of thieves who had posed as buyers. He used the Find My iPhone app to track them down. (Police had to save the would-be thieves from an angry mob that had gathered to watch the takedown.)

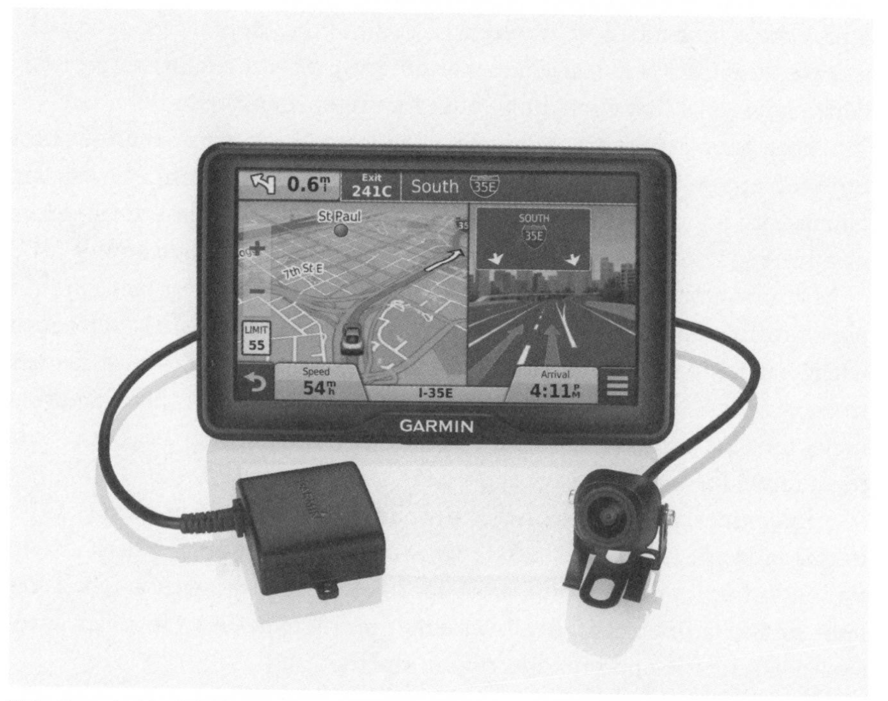

This Garmin Nuvi GPS system with backup camera is typical of standalone GPS units used in vehicles. *Photo courtesy of Garmin, Inc.*

Of course, if a phone's GPS functionality alone weren't enough to catch hapless thieves, some of them are dumb enough to compound their idiocy by taking celebratory (or occasionally taunting) selfies: photos of themselves shot with the very same smartphones they had just nabbed. That way, the police not only know where they are, they even know what they look like. Crime probably does pay, at least for some, but not for folks dumb enough to steal items that can pinpoint the locations of both the thieves and their booty—and certainly not for crooks who stop to take photos of themselves, photos that are often immediately and automatically e-mailed or posted to the victims' Cloud-based photo albums.

A Little Common Sense, People

On the other hand, blindly trusting a technology is a sure recipe for disaster, and GPS is no exception, as many can attest to—especially those who unthinkingly follow GPS instructions and find themselves in the middle of a pasture or, worse still, at the edge of a cliff. (One hopes. If they actually followed the GPS unit's instructions and drove *off* of the cliff, we probably don't know *what* they thought because they're now unable to tell us.)

Take, for instance, truck drivers trying to make it across the hills near Stowe, Vermont. One convenient route is what locals call "Smugglers' Notch." Taking it cuts hours off of a cross-mountain trip, but not if you're in a big-rig truck. The road is not open for truck traffic, but not all GPS units know that. The nice, somewhat robotic-sounding lady (or man) in your phone or freestanding GPS unit will gladly guide you into The Notch, but when you get to the top, a hairpin turn and a large rock wait to snag unwary truck drivers. GPS-guided big rigs routinely get stuck up there, shutting down the road and costing the state (and the truckers) thousands of dollars by the time the mess is cleaned up.

Sometimes, though, you have to wonder if more than just faulty GPS info is at work. Last winter, a Belgian woman attempting to get to a train station just north of her home drove for hours (stopping twice to refuel and once to take a brief nap), finally ending up in Croatia—1,450 miles away, and in exactly the opposite direction of the train station.

"I was distracted," she told authorities.

Then there's the New Jersey family that came upon a T intersection (where, if you're paying attention at all, the only options are to turn either right or left) and instead opted to follow their GPS's instructions—which were to continue straight ahead. The driver ended up running over the lip of the opposite curb, but even then did not stop until he'd hit a house about a hundred feet away.

Of course, none of us are immune. A few years ago my wife and I were in Sacramento visiting a friend. Late at night we headed back to our hotel—or so we thought. I had entered the state, city, and street name into our stand-alone GPS unit when, without waiting for the full address, it displayed a route. Being male and foolishly unconcerned with directions (two characteristics that, according to my wife, are essentially identical), I simply hit the GO button. That was mistake. It turns out that the street for which we were headed is a *very* long one. Miles long, in fact; it stretches from one end of town to the other. The section on which the hotel was located is actually quite nice: well-lit, tree-lined, trendy restaurants, cute little cafes and pubs. You get the picture. Unfortunately, the section of the street on which *we* ended up was a dark and sketchy industrial area, and more than a bit unnerving. It took quite a while to make our way to the other side of town, during which time my wife might have commented once or twice (or so) on my unerring lack of a sense of direction.

I could go on with example after example, but the lesson is clear: Use technology, but also use your common sense. And if you're directionally challenged, listen to your spouse.

Using GPS to Do Science

Well, not *your* GPS, of course. But in the hands of experts, these handy devices can do more than help you find the nearest sushi bar.

Weighing Water

Researchers from UC San Diego and the US Geological Survey have used GPS to study drought in the western United States. It turns out that water is pretty heavy (about eight pounds per gallon), and when a section of earth is relieved of some of that weight, it tends to rise; not surprisingly, scientists call that movement "uplift," and it can be measured. And since GPS measures movements of all sorts—including vertical movement—it turns out that one can insert a sensitive GPS unit underground and use it to calculate the amount of water that's missing by measuring the amount of uplift that results when the water disappears. (In the event, scientists reported that about 240 gigatons (Gt) of water had been lost over the past eleven years. That, by the way, is enough to cover the entire region in a four-inch layer of water.)

Examining Atmosphere

If that's not impressive enough, scientists also use GPS to examine our atmosphere. Special lower-altitude GPS-radio occultation (GPS-RO) satellites work in conjunction with the standard GPS satellites: As the standard satellites pass below the horizon of Earth (relative to a GPS-RO satellite), the radio waves emitted by the GPS satellite bend and slow down as they pass through the atmosphere; how *much* they bend and slow down before being received by the GPS-RO satellites can tell us much about the composition of the atmosphere through which they are passing, including its temperature and humidity.

Tracking Shifting Tectonic Plates

Mountains move. Entire continents move. When they begin to move in an erratic fashion, it can be a sign that an earthquake is imminent. In central Washington State, scientists have set up PANGA: the Pacific Northwest Geodetic Array. PANGA is a network of GPS sensors that can detect even minuscule movements in order to calculate the buildup of strain on various earthquake faults. Someday, PANGA may help us to better understand the forces at work on our planet—or even allow us to predict earthquakes with stunning accuracy.

Fluffy: The Stone-Cold Killer

What happens when you put GPS collars on supposedly domestic cats? You discover that they're actually not so domestic. They like to kill small mammals and birds—some of which they bring back as offerings to the people with whom they grudgingly share their homes. Interestingly, they also tend to "timeshare"—restricting their wanderings in certain areas so as to miss other cats, thus avoiding territorial disputes (although a few appear to actively seek out opportunities to fight).

That's what researchers Alan Wilson of the Royal Veterinary College, Sarah Ellis of the University of Lincoln, and John Bradshaw of Bristol University discovered in 2013 when they put tracking collars on a group of fifty domestic felines. (They'd done the same thing with cheetahs in Botswana, but decided to try the same technique on little Fluffy because . . . well, because no one ever had, and because we've all wondered what the heck it is that Fluffy does all night.)

"In fact, we know less about some aspects of their behavior than we do about many wild cats," said Dr. Wilson in a *BBC News* article. "So the Horizon program and the study in our chosen village . . . was a fantastic opportunity to find out some of this missing information."

Among other discoveries, a few of the less grisly: Domestic cats normally don't wander as far as one would think. In a small English village surrounded by countryside, most stayed within the village itself, roaming perhaps six hundred feet or so from their homes. They also like to visit the homes of other cats, often going inside to interact with the people who live there—and, one assumes, with their feline friends.

By the way, if you think Fluffy is a killer, wait 'til you see the cheetah data recorded by Dr. Wilson and his colleagues in Botswana. Cheetahs really *are* the world's fastest land creature, easily outrunning Olympic-class sprinters, and even outperforming racehorses. The cheetahs' hunts were successful an astounding 25 percent of the time (much better than, say, a lion might do), even when they didn't really exert themselves, and they seemed to hunt as much at night as during the day. Also, their speed (and hunting success) remained about the same regardless of whether they were running on flat, empty land, or through dense vegetation. So, if your first thought is to outrun a hungry cheetah, you really need to come up with a better plan.

Improving People's Lives

GPS is more than merely a tool (however impressive) that you can use to help find your way to a theater or to that B&B you've been wanting to check out. In the hands of concerned citizens, governments, and organizations, the technology is being used as a force for good.

One example: Over fifty million people live in the Congo Basin. Of that number, somewhere between 500,000 and 700,000 are indigenous natives living in the rain forest and comprising some twenty-six distinct ethnic groups, each with its own language and its own customs. They are hidden away in the depths of an almost-impenetrable lushness, often marginalized or simply ignored by those in power: politicians, developers, builders, and others. Bowing to the forces of industrialization and population growth, the indigenous people have been powerless, able to do nothing more than watch as the forest—their home and their livelihood—has been taken by developers and industrialists, leaving both the people and the game they hunted homeless.

"Perhaps the biggest challenge we face," says anthropologist Joe Eisen, "is changing attitudes of certain governments and donors that the existing paradigm of forest governance—that of promoting industrialization of the forest while creating conservation 'islands' where local people are forbidden—is not a viable long-term solution to poverty reduction or conservation of the rain forest."

Eisen is the coordinator of Rainforest Foundation UK's geospatial mapping program, Mapping for Rights. The project uses GPS as a tool for charting the forests to show regional and national governments that the rain forest is not simply an empty, unused wasteland: GPS-created maps show that *people* live there, and that they have rights that must be respected. They build villages, they hunt and travel well-marked trails, they visit residents and relatives in other villages; the forest is their home, and it is rapidly disappearing. Since 2000, Mapping for Rights has been working to create maps as a way of improving governance of the tropical rain forests and of promoting the rights of forest communities.

It's more than a little ironic, actually: Mapping—so long a tool of empire—is now being used in defense of people disenfranchised by that empire.

"Cartography has been used over the centuries as a tool to create and perpetuate power, property ownership, and political claims, presenting a vision of territories as perceived by dominant or official powers," says Eisen. "In recent years, participatory mapping has emerged as an alternative to empower marginalized groups and those traditionally excluded from decision-making processes, such as rural and indigenous communities, to integrate their spatial knowledge and worldviews onto maps and contributing to a reversal in power relations."

For Rainforest Foundation UK and its partners, a key part of a possible solution is technology that can be used to show that these indigenous communities actually exist and have claims to the forest.

The organization's Mapping for Rights project has been helping indigenous peoples to secure their rights "in the face of unprecedented pressure on the forest from the extractive industries and others," notes Eisen. "Our mapping work has gradually evolved from using fairly rudimentary techniques, such as compasses, to capitalizing on new geospatial technologies based on GPS and GIS systems as these have become more available, helping to improve the precision and scale of the mapping work."

Today, the organization's mapping program may be the largest of its kind in Africa, having covered over two million hectares to date (a hectare is a little less than two and a half acres), with plans to map a further five million hectares over the next three years.

As technology continues to evolve, so does the group's use of it.

"Future mapping," says Eisen, "will be done with tablets using icon-based open-source software that allows even illiterate people in remote forest areas to collect precise geo-referenced data about their customary lands and development needs and priorities. Data collected is then uploaded to an interactive web map, where it can be accessed and analyzed by policy makers in the context of other land uses, such as logging concessions and mining permits, to make more-informed decisions about the forest and its inhabitants."

What's In Store for GPS?

So, what's next for this technology? To what uses might it be put? While no one can predict the future, there's certainly something to be said for looking at existing technologies and extrapolating—imagining how they could be used, where they might take us, and how they might help to solve pressing problems. We can be pretty sure that the granularity—the resolution and accuracy of GPS devices—will continue to improve, but what about beyond that? What uses will we make of this increasingly accurate tool?

Most of us lack the creativity necessary to think "outside the box" in a way that can lead to sudden flashes of inspiration. It takes quite an imagination, and not all of us possess that special faculty that allows us to break free of our preconceived notions of how a technology might be used.

But Jorge Treviño has that imagination, and the talent to back it up. Treviño is a young industrial designer in Monterrey, Mexico. He specializes in designing things that are both useful and beautiful. I asked him which was most important to him, form or function—and, like the intelligent young man that he is, he hedged a bit:

> *I like to explain my opinion by analyzing the umbrella. It has its origins in China around four thousand years ago. We still use it, and yet it hasn't changed much. What has made the umbrella so useful has been its function of covering the user from the rain, but that function is followed by its form.*
>
> *Form and function are both dependent on each other, and when they are in perfect harmony, you achieve good design.*
>
> *If we design a smartphone for the military, one intended for use outdoors and on the battlefield, it would have to be durable, designed to last, and have an extremely good signal all the time. In this case the slickness of the smartphone, or the chrome on its bezel, is not going to be as important as it is when designed for the masses. Some other features of this product are going to have to be more predominant, and the form of it has to be the perfect one for the smartphone to achieve its purpose. That's the real beauty of an object.*
>
> *However, sometimes the function of an object, or its purpose, is determined after its form. This can happen when experimenting with different design methodologies. After a creative process, an interesting form can be conceived, and one may later realize that it's a perfect way to solve a specific problem.*

Treviño has come up with a design for a GPS-enabled tool that could alter the way sight-impaired people find their way around—in fact, it could even alter the way they live their lives.

The tool is called Discover, and it's a handheld device that uses pin-screen impression technology combined with a camera to create tactile 3D maps of a user's surroundings. The user can *feel* what the device's camera sees by touching the pins, which use the onboard camera to arrange themselves into a 3D representation of one's surroundings.

"It uses a camera that detects movement and objects in front of it," says Treviño. "It then translates the image into tangible surfaces that a user can feel with his thumbs and fingers."

The device looks like a cross between a *Star Trek* communicator and a futuristic remote control. A rectangular pin surface on the upper part of the

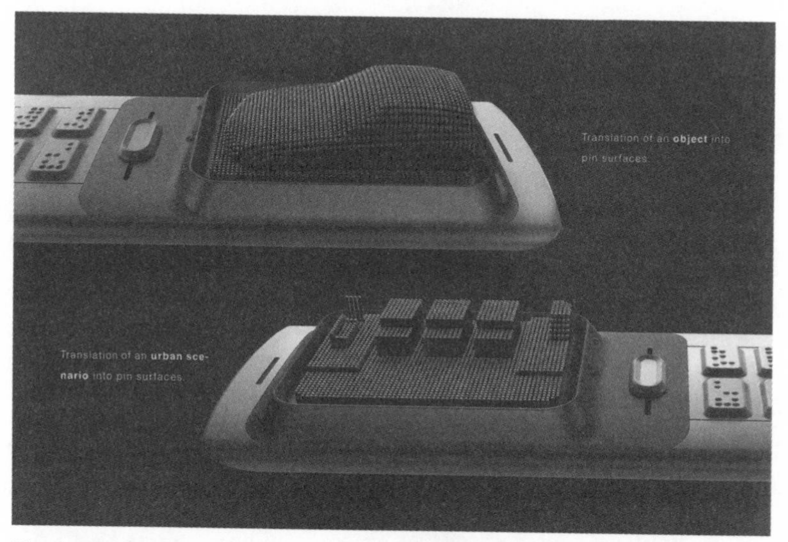

Discover translates a scene into 3D pin surfaces that can be touched by a sight-impaired user. *Image courtesy of Jorge Treviño.*

top side of the gadget displays the scene recorded by the camera. Below that are six buttons, labeled (in Braille) "Discover," "City," "GPS," "Read," "Time," and "Tag." On the front is a camera that records the scene to be displayed on the pin surface. The design for the underside of the device calls for a matte finish that, along with a wrist strap, helps the user to keep his grip on the device.

Each button corresponds to a mode of use that can be selected by the user:

1. *Discover*: This is the standard mode, in which you can walk around and discover the environment that's in front of you. Through the different scales, you can zoom in and out on the images.
2. *City*: In this option, Discover uses GPS to identify the streets near where you are walking. With a 3D map, you can *feel* the streets on the map and know where you are.
3. *GPS*: With this option you can say your destination and, similar to the "City" option, Discover will guide you through the map to reach your destination.

4. *Read*: This mode shows Braille text on the screen so that the user can read by scanning text or by downloading an e-book.
5. *Time*: This is a simple mode that shows the time on the 3D screen.
6. *Tag*: This option uses GPS as well. By pressing this button, Discover saves your current location. You can name the place and then navigate to it later by saying the name out loud. If the user wants to know what's in front of him but can't understand the surfaces on the screen, he can press the pins on the screen and Discover will tell him the name of the object.

"My goal was to help people with this disability increase their freedom," says Treviño. "Walking on the streets they can easily get lost, and unfortunately, most of the streets weren't designed with sight-impaired people in

Inventor/designer Jorge Treviño came up with the idea for Discover, and is seeking to make it available through partnerships with manufacturers. *Image courtesy of Jorge Treviño.*

mind. Hopefully, with Discover they [can] move around, identify what's going on in front of them, and avoid accidents. The screen not only displays representations of imagery; it can also display Braille whenever the user prefers," says Treviño.

A conceptual project based on an MIT Tangible Media Group technology, Discover is not yet a real product. However, Treviño is hopeful that he'll eventually be able to partner with government or private industry to bring it to market. If and when he's able to do so, it will be yet another testament to imagination, and yet another example of a new technology being used to improve people's lives.

The Nuts and Bolts of GPS

So all of this about the uses of GPS is very interesting, but how does it work, anyway? It seems magical, doesn't it? And the first few times you use a GPS-enabled device or application, it feels like some form of wizardry. Somehow, this piece of equipment knows *exactly* where you are, within a couple of feet. (Actually, it knows exactly where *it* is; you just happen to be nearby.)

But it's not magic, it's something even more amazing: It's math. Geometry, mostly, but with some other flavors of math thrown in for good measure. (Including calculus, which happens to be very good for studying and predicting how things change—things such as the locations of orbiting bodies.)

About twelve thousand miles above the Earth is a constellation of specialized GPS satellites. Some of the satellites orbit the planet every two hours, some take as long as twenty-four hours. Because of the number of satellites (the designers of the system anticipated that a total of twenty-four would constitute the entire network, but sometimes there are more, and there are always a few ready to perform backup duties), there are always some satellites orbiting above wherever you happen to be on Earth. At press time, there are thirty-one working GPS satellites in orbit. Note that, in spite of what you may have heard, the satellites are *not* in geosynchronous or geostationary orbits—orbits in which the orbiting body remains fixed relative to the Earth below it due to its matching the Earth's rotation, or else remains fixed above the Earth's equator. This makes the math used to triangulate a position a bit more complicated, since the satellites' locations constantly change relative to Earth, but it's nothing we can't handle in real time with computers.

Aboard each satellite is a very accurate clock. Each satellite transmits signal information to Earth, and your handy GPS receiver takes this information and triangulates to calculate your location. Contrary to popular belief, the GPS device itself does not transmit, so the satellite does not know where you are; thus, you cannot be tracked by a satellite (tinfoil-hat theories notwithstanding).

And here's why the clock is important: Each signal is time-stamped; it carries with it metadata that includes, among other things, the exact time it was transmitted, based on the super-accurate clock aboard the satellite. Your receiver compares the time a signal was transmitted to the time it was received; the difference in time—that is, how long it took the signal to reach your receiver—tells the GPS receiver how far away the satellite is. Given measurements from multiple satellites (at least three are necessary to get an accurate fix), the receiver can determine and display your position on the map displayed on the unit's screen.

It's not magic, but it certainly is impressive.

The Longitude Problem

We can hear an interesting and oddly appropriate historical echo here, one that goes back to the Lewis and Clark Expedition clock, and then back hundreds of years before that.

Seafarers in the eighteenth century had a problem. Well, they had lots of problems, actually, including scurvy, floggings, pirates, keelhauling, drunkenness, crews that had been unwillingly "Shanghaied" aboard, and the strange fact that few men aboard merchant vessels in those days could swim. But the most vexing problem of all was how to find one's way on a voyage. Specifically, how to determine one's longitude.

Latitude, you see, could be fairly easily determined by sighting the sun at noon and using accompanying mathematical tables that indicated that if the sun was at such-and-such a position at noon (i.e., at its highest point, whatever the actual time might be), then your latitude was so-and-so. It was a little tricky and required some skill with a sextant, but normally something that could be done. (Though problematic at night or on a very cloudy day, of course.)

But longitude—your exact distance east or west of Greenwich, England (or any other reference point)—well, that was a different story. Some of the best minds in the world had been laboring for hundreds of years to find a simple and accurate way to determine longitude while at sea. It was a very serious issue, and many lives (and fortunes) depended on a solution. In fact, so much was at stake that in 1714 the British government offered a £20,000 Longitude Prize, equivalent to millions of today's dollars, to whoever could solve the problem by finding a way for ships at sea to determine their longitude. (The kings of France and Spain offered similar prizes.)

The answer turned out to be a clock. Specifically, a clock that was so accurate and so robust that it *always* indicated the correct time at some point of reference. Plenty of clocks existed that were accurate enough on land, but creating one that could keep time accurately on a moving ship was thought by many to be impossible.

John Harrison knew better. The brilliant English clockmaker was sure that he could build a clock—a chronometer—that would be unaffected by wild variations in temperature and humidity, and which could keep reliable time, even on a moving, rolling ship. And he was right, although it took him forty years to develop the timepiece—in essence, an oversized and amazingly accurate pocket watch—known as "H4."

However, mostly for political reasons, Harrison was repeatedly denied his prize. In the end, King George III had to weigh in on Harrison's behalf, and even then Harrison never received the official prize, though he was paid in increments that eventually amounted to just about its full value, or a bit more. He died a wealthy man, and deservedly so.

Ironically, but not at all coincidentally, the mechanism that allows a GPS to correctly display its location and the solution to the historical longitude problem both hinged on the same thing: an accurate clock. Or, in the case of GPS, twenty-four clocks.

A Brief History of GPS

Like many technological developments, GPS owes much of the impetus to its development to war—or, more accurately, to the *threat* of war.

In 1955, some two hundred years after Harrison's chronometer, a physicist named Roger Easton was working for the US Naval Research Laboratory. Easton worked on the NRL's Vanguard proposal as part of a competition to determine who would build America's first satellite, meant to be launched as our opening salvo in a Cold War competition that the Soviet

Union seemed at the time to be winning. (Vanguard's US Army–sponsored competition included a proposal by the army's pet scientist, Wernher von Braun, who had been spirited out of Germany after the collapse of the Nazi regime.) Project Vanguard won the competition, and in 1957, the year *Sputnik* was launched, Easton invented a tracking system to help determine the satellite's orbit. That system, Minitrack, continued to be enhanced and updated over the next few years.

In 1959, Easton and his team designed the Naval Space Surveillance (NAVSPASUR) system, the first system able to detect and track *all* types of Earth-orbiting objects. Through the 1960s, Easton and others worked on what would eventually become TIMATION, a system of satellite range-trackers based on the idea that each satellite would have on board one or more super-accurate clocks used to help determine the satellite's location.

In the 1970s, the indefatigable Easton led the development of time-based technologies that would eventually be used to build the United States Global Positioning System, and in 1977, the navy's NTS-2 satellite was the first satellite to transmit GPS signals. As well as being the first satellite to use a nickel hydrogen battery, NTS-2 had on board two atomic clocks; the launch was a test flight meant to show the efficacy of satellite navigation based on precise timing, and it succeeded handsomely.

Property of the US Government

The result of years of secretive military research (and boasting undeniable military application), our GPS system is actually owned and operated by the United States government as a national resource, under the stewardship of the Department of Defense (DoD). An Interagency GPS Executive Board (IGEB) oversaw GPS policy matters from 1996 to 2004. In 2004, the National Space-Based Positioning, Navigation and Timing Executive Committee was established to advise and coordinate federal departments and agencies on matters concerning GPS.

From the mid-1970s on, GPS navigation technologies were solely the purview of the US military. Civilian use was simply against the law—even assuming that one could figure out how to build something that could receive classified data transmitted from satellites orbiting many thousands of miles away. GPS was a military technology, period. Thinking that a civilian might one day have access to it was roughly akin to believing that you would someday be able to go downtown and buy an Abrams tank.

That changed in 1983, when, as already noted, Korean Air Lines Flight 007 was shot down after straying into Soviet airspace. In the wake of this disaster, President Ronald Reagan issued a directive making GPS freely available for civilian use, once the system was fully operational. (Having decided on that course of action, it took only five years from the launch of the first official GPS satellite in 1989 to the launch of the twenty-fourth.)

And yet, even after the presidential directive, and even after dozens of electronics manufacturers had entered the GPS receiver market, "freely available" turned out to be subject to a certain amount of interpretation. Allowing civilians access to a closely held military technology such as GPS did not necessarily mean that the government was ready to give *complete* access.

For years, the government limited not access to the tool itself, but the *accuracy* of the tool. While military GPS was accurate to within a few inches, civilian GPS accuracy was limited to within several feet—still more than close enough for casual navigation, but nowhere near the pinpoint accuracy of military applications.

In May 2000, President Clinton announced that one civilian GPS limitation, called "selective accuracy" (or SA), was to be immediately removed; a few years later, the government permanently eliminated SA by mandating that future GPS satellites be built without that capability.

These days, civilian GPS provides "worst-case" accuracy of 7.8 meters, or about 25 feet, and often quite a bit better than that. (And augmented by other technologies, including cell-tower triangulation and the RTK-enabled devices described above, one can pinpoint a location within a few feet, or even a few inches.) Currently, military users of GPS can take advantage of "ionospheric correction," while civilian users cannot. (Ionospheric correction takes into account any delays introduced by the ionosphere as the signal encounters and then passes through that layer of the atmosphere.) However, the US Air Force notes, "Eventually, the accuracy difference between military and civilian GPS will disappear. But military GPS will continue to provide important advantages in terms of enhanced security and jam resistance."

So, even when the technology is *completely* opened up to civilian use, "completely" is (possibly for very good reasons) also subject to interpretation.

Operators at the master control station at Schriever Air Force Base track orbiting satellites. *Image courtesy of United States Government.*

In the end, GPS is an excellent example of one of the primary theses of this book: Technologies tend to start out closely held, with few citizens having access to them. In this case, GPS had been available experimentally since the 1970s, and then in the 1980s became a technology freely available to civilians. By the time the system was fully functional, civilians had unfettered access to the signals themselves, but civilian signal reception had been "degraded" to a point at which military users could enjoy a significant advantage in accuracy.

Now, even that advantage has (at least as far as we know) been eliminated, allowing manufacturers, developers, experimenters, and entrepreneurs to use the technology to its fullest, and to create, discover, and innovate along the way.

When the rest of us get our hands on these new tools, an explosion of innovation and creativity results. Witness the hundreds of GPS-enabled devices, from dash-top receivers to smartphone navigation apps, and from fleet-management programs to projects meant to save the environment.

According to some estimates, the market for such tools will be worth over $26 billion by 2016, making this one case where a few billion dollars in taxpayer-funded research and development turns out to have been well worth the cost.

"It's one of the rare government projects that you can definitely say we've gotten bang for our buck," said Richard Easton in a *Business Insider* article. Easton says that GPS expands our GNP by as much as $100 billion a year, while costing only $1 billion in maintenance. Easton, a lecturer and writer, is the author of *GPS Declassified: From Smart Bombs to Smartphones*. He is also, as you may have surmised, the son of physicist and GPS inventor Roger Easton.

GPS is an example of technology that was first withheld from civilian use and then, when civilian use *was* allowed, one the effectiveness of which was purposely degraded. Now that it's in common use and tools for its development and integration are widely available, we find that it spurs further development, provides jobs, and benefits the economy—not to mention the fact that it can be used to help disenfranchised people, speed up the delivery of emergency services, and conserve resources. All in all, a solid investment, and a good argument for getting technology into people's hands as quickly as reasonably possible.

The development of GPS created an entire industry, and may yet lead to more geo-aware tools and to unforeseen further developments. Another new technology may not only create a new industry, but could also disrupt several existing ones. The net result of that disruption may well be viewed favorably in the end, but any disruption is painful and disorderly, and this one may lead to loss of jobs and to the need to make radical alterations to industries that have remained largely unchanged for generations.

The new development is 3D printing, and it is to that technology that we turn next.

Chapter Eight

A Revolution in a Box: 3D Printing

Almost everyone who has had an idea that's somewhat revolutionary or wildly successful was first told they're insane.

—Larry Page

Your Next New Home May Have Just Gotten Much Less Expensive

If you set out in 2013 to build a new home, you would have paid, on average, about $400,000, including land and finished lot costs, such as public utilities, platting, auditing, etc. The land alone would have accounted for about 19 percent of the total. Construction costs would have amounted to 62 percent. Your new home would have taken several months to build. (By the way, only about 9 percent or so would have been builder profit, so don't think your builder is necessarily getting rich off of you; if he *is* wealthy, it almost certainly took him many years and a great deal of hard work to get that way.)

But what if things were different? What if you could save, say, 25 percent on materials? And maybe cut your labor costs in half? And what if, instead of taking several months to build, your new home could be ready in, oh . . . twenty hours or so?

It's not as crazy as it sounds. The newest and most radical up-and-coming construction technique has little to do with hammers and framing and everything to do with computers and printing—specifically, 3D printing.

The Origins of 3D Printing

As noted earlier, 3D printing was the 1983 brainchild of a middle-aged engineer named Chuck Hull who was looking for a way to quickly and inexpensively prototype plastic parts. At the time, designing and fabricating a prototype part took weeks or months, and could cost a small fortune—and in the end, the part might in fact turn out to be useless.

Chuck had an inspiration, and it was nothing short of brilliant. In fact, it was one of those simple but magnificent flashes of insight that makes the rest of us wonder why *we* never thought of it. He realized that one could build a part in layers—a process that we now call *additive manufacturing*—by spraying a thin layer of material in the desired shape, and then spraying more layers on top of that first one, building up the item one layer at a time and allowing each layer to dry until one had a fully formed part. The machine doing the spraying would be controlled by a computer that sent instructions to a mechanism riding on a carriage upon which was mounted a "printhead" of sorts, one that dispensed not ink, but liquefied plastic or metal. The carriage itself (or the base on which the printed item was being created) could rise or fall as needed to accommodate the layers being added and the shape being built.

Along with the printer, Chuck had to design software that translated the part's specifications and fed them to the printer, and also the file format that the software read in order to begin translating the item specs into instructions it could send to the printer.

In the end, Hull had created a revolution. Parts could now be designed, prototyped, and tested in hours instead of weeks. Eventually, companies using large-scale 3D printers that created parts of metal or tough composites began building not prototypes but actual parts. (In April of 2011, Boeing launched the ninety-foot Phantom Ray drone, a UAV with a fifty-foot wingspan and a top speed of over 600 miles per hour. The drone was designed, built, tested, and deployed within two years, largely because a huge percentage of its components were 3D-printed.)

But it's not just governments and huge corporations benefiting from 3D printing these days. Now 3D printers are showing up at your local big-box electronics and office supply stores. Even e-commerce titan Amazon.com (which started out selling books and now sells just about everything) has a

"3D Printing Store" section highlighted with a "Shop the Future" banner, in which one can purchase a 3D printer for as little as $500.

Only it's not the future; it's the present.

"While the average consumer currently has little use for a 3D printer, these machines could potentially democratize manufacturing, much as the Apple II did for computers," said technology writer Tyler Wells Lynch in a 2011 *USA Today* article. "Homeowners might be able to print mugs, doorknobs, jacket hooks, and other household items on a whim."

A few years ago, when Lynch wrote those words, 3D printing was just beginning to be discussed in consumer circles, and no one was sure that it would have much of an economic impact. But this year, *Business Today* estimates that within ten years, 3D printing will be an $8.4 billion market. And its impact will reach far beyond mugs, doorknobs, and whimsy.

RepRap: Bringing the Revolution to the People

We spoke of Dr. Adrian Bowyer's revolutionary RepRap 3D printer in the introduction. As you'll recall, Dr. Bowyer, formerly a mathematician and lecturer at the University of Bath, looks at 3D printing as something that could revolutionize manufacturing. If everyone could have a 3D printer, he reasons, think of what could be accomplished; think of the creativity that could be unleashed, and imagine what could happen when just about anyone could manufacture just about any item he needs for a few pennies and a little time spent on a computer.

Dr. Bowyer is not sitting back waiting for that revolution to occur. He's nudged it along—perhaps violently shoved it forward—by inventing RepRap, an affordable open-source 3D printer that is essentially self-replicating; that is, it's meant to be used to create more instances of itself. RepRap is a printer that can create more printers. Dr. Bowyer thus envisions a day in which anyone who wants a 3D printer can have one.

But he didn't start out to revolutionize manufacturing.

"I have always been interested in the idea of making a useful self-replicating machine. When my department got a grant for equipment, I was responsible for spending some of it, and I bought two 3D printers. I realized that finally humanity had a manufacturing technology sufficiently powerful to stand a chance of self-replication."

This RepRap printer, "ORDbot quantum," was created by Bart Dring. *Image licensed under GNU Free Documentation License 1.2 via Wikimedia Commons.*

The power of Dr. Bowyer's idea was obvious, though: Once people begin creating things, once they begin *manufacturing* items in their homes or dens or garages, there might be no end to the creativity and innovation that could result.

How ubiquitous would Dr. Bowyer like to see 3D printing become? He's thinking about near-universal use of the technology: "I would like to see the same proportion of people in the world making things as currently use text editors in all their various forms—from Facebook-entry editing to writing a novel using a word processor to typing e-mails. To get that large a number of people making things for themselves would be what I would want."

That seems like a revolutionary idea—literally revolutionary. There's no hyperbole here, either; this could upset some basic tenets of capitalism. (One 2006 article said that RepRap has been called "the invention that will bring down global capitalism," which might be a bit of a stretch. But turn *some* things upside down? Definitely.)

Imagine a world in which if you need a bathroom hook—or perhaps someday, an entire bathroom—you could simply print yourself one. That sort of power (and we are, after all, speaking here about how technology can empower people) is disruptive, to say the least. It means that our traditional notions of manufacturing, warehousing, shipping, and delivery are, at least in many cases, out the window. (A window whose frame and sash you may have in fact printed in your home.)

But Dr. Bowyer thinks that such a revolution would ultimately be a good thing, noting that the best thing about such tools is "the enormous scope they give to invent new devices and pieces of software and to see if they work."

What he's talking about is no less than a tool that can unleash the power of design, of creativity—in fact, the power of *creation* in both a metaphorical and literal sense. That creative power, he says, must be shared by, and available to, everyone.

"*Engineering* is a fancy word for making things," says Dr. Bowyer. "And every nation that has embraced engineering in the last few hundred years has made its inhabitants wealthy and comfortable beyond the dreams of their preindustrial ancestors. The ability of *everyone* to make things can only enhance this."

The Growth of 3D Printing

With or without RepRap, 3D printing *is* becoming more prevalent, and it's morphing as we watch, changing from a tool used to build prototypes into a ubiquitous technology used to create actual finished products. It's a technology revolution, and it's happening before our very eyes. Not interested in actually owning a 3D printer? You can instead choose to make use of service agencies that provide access to the machines.

Take New York–based Shapeways, for instance. It's a full-service 3D printing company founded in 2007 that can create objects in a variety of media, including plastic and metal. In fact, Shapeways can print your item

This golden key was printed by Shapeways. *Photo courtesy of Shapeways.*

in any of forty different materials, including gold, brass, stainless steel, and silver. Just use CAD/CAM software on your home or office computer to create a design and then upload it to the Shapeways website. Just a buyer, not really a maker? You can purchase other people's 3D creations in Shapeways's online store.

The ultimate result of the proliferation of such services? The eventual localization of manufacturing. As mentioned in the introduction, this is not a new idea at all. It's exactly how guilds and workshops operated for centuries, with everything from horseshoes to jewelry being sold at the place of manufacture.

That's a possibility that Shapeways CEO Peter Weijmarshausen endorses wholeheartedly. At the 2014 TechCrunch Disrupt conference, he noted that manufacturing *should* be local. "Our view is, over time, more and more Shapeways factories will appear in places all over the world, and in those places, not only will we get products to you faster and at a lower cost, but we'll have lower impact on the environment for transportation."

Another provider, one with a different business model, is makexyz.com, which is an interesting example of a service economy based on a new technology. Makexyz doesn't actually *print* anything. In effect, the company acts as a broker, connecting people who need 3D printing with someone in their area who has a printer and is willing (for a fee) to print the object requested.

And, as is typical of many innovations, makexyz was born out of frustration.

In January of 2013, Nathan Tone and Chad Masso, two recent college graduates, were working for Indeed.com, an Internet-based employment search engine. The two had come up with an idea for a simple light-switch cover that had a hook on the bottom, so that one could walk in the door, flip on the lights, and leave a set of keys on the hook.

But Tone's 3D printer was out of service at the time, forcing the two inventors to track down and contract with a centralized service in New York.

"It was expensive," says Tone, "and it took three to four weeks to get the part. Which was frustrating, because there was certainly someone with a printer near us; we just didn't know who or where they were."

Out of that experience came makexyz, a company that matches up people with ideas to people nearby who have compatible printers.

People like Clay Cardwell, for instance. Cardwell is an Omaha-based computer service representative who in his spare time prints 3D movie props, mostly as a hobby. These can be as simple as a small "thermal detonator" (as seen in *Star Wars*) or as complex as an entire Iron Man costume, but all of them come from the 3D printers in his home workshop.

Cardwell is a makexyz vendor, and his take on the service is straightforward: "Simply put, you create an account and let people know what device you have and are willing to use for making things—like a 3D printer, a CNC machine, or a 3D scanner—and you indicate the basic price you'll charge. With my 3D printer, I charge twenty-five cents per cubic centimeter of plastic. So people send me inquiries and if I think I can print what they need, I accept the job and send them an invoice. The site [makexyz] makes a small percentage, and I get paid for the work."

Of course, with 3D printers becoming more and more common, won't there eventually come a time at which the market is saturated? Cardwell and Tone don't appear to be worried about that.

"I don't see it reaching a point of saturation for a long time," says Cardwell. "What I do see it doing is opening the door for creative people to come up with unique uses. For instance, there's a gentleman who is 3D-printing artificial hands for children with birth defects. He collaborated with another

A 3D-printed cup with the *Computer Power User* magazine logo as a handle. *Photo courtesy of Andrew Leibman.*

person halfway around the world, and he builds them for free. He even put the plans out on the Internet on Thingiverse.com, which is a website dedicated to open sharing of 3D files. I have stuff out there myself."

Tone also doesn't see demand slacking off for various 3D print services, even with the increased popularity of such printers. "The biggest challenge today is that only a tiny slice of the population understands what's possible with 3D printing. And if 3D printing achieves a 'saturation point,' it has also by definition achieved a massive market, one in which the overwhelming majority understands and needs 3D printing," says Tone. "In that scenario, makexyz thrives; maybe everyone has a basic printer, and then you go to makexyz to print your electronics, etc."

"I don't see [increased popularity] as a bad thing," says Cardwell. "I hope it leads to more devices like the 3D printer to help improve life for mankind. I would love to see the day you buy a house with a built-in *Star Trek*–like replicator. NASA is adapting 3D-printing technology for space. And when we make it to Mars we'll need a device like the 3D printer to quickly build stuff, since we can't carry a lot of materials there."

Cardwell may be on to something here. NASA has actually been experimenting with the use of 3D printers in space, and for exactly the reasons that Cardwell notes: Sending a replacement part from Earth to the International Space Station is an expensive and time-consuming proposition. (It can take months, in fact. Sending parts to Mars would take years.) But, given a 3D printer and a supply of raw materials, astronauts could easily manufacture their own parts, even one-off parts of their own design. That would not only save time and money, but it could at some point save lives.

Made In Space CEO Aaron Kemmer described the advantages of extraterrestrial 3D-printed parts in a CBS News interview in September of 2014: "With 3D-printing technology, you can just get around that entire [conventional manufacturing] paradigm and essentially *e-mail* your hardware to space." NASA's 3D-printing experiment envisions printing not only parts, but also the tools used to assemble those parts. The first space-bound 3D printer was launched in September of 2014, carried aloft by another sign of the democratization of technology: a privately owned SpaceX Falcon rocket. Three months later, astronaut Barry Wilmore—

The Made In Space 3D printer was launched into orbit in September 2014. *Photo courtesy of Made In Space.*

needing a ratcheting socket wrench to complete a repair—simply *printed* one after NASA e-mailed him the plans.

Prosthetics, Exoskeletons, and . . . Organs? Seriously?

It's obvious that 3D printing could turn conventional manufacturing on its head. There will come a time when just about anyone has the ability to create everything from key chains to clothing, and from dolls to dinettes. The real question is, with such a broad spectrum of possibilities, can we determine where the technology's *greatest* impact might lie?

Cardwell may have hit upon it when he alluded to the profusion of printed assistive devices. The health-care profession has shown great interest in the possibilities of 3D printing. With printers now being used to create everything from orthotic shoes to custom-fit casts, there seems to be no limit to how far medical innovators and researchers can take the 3D-printing revolution.

In 2013, Paul McCarthy embarked on a search for a prosthetic for his young son, Leon, who was born without fingers on his left hand. McCarthy came across a website and videos hosted by Washington state inventor Ivan Owen. Owen, who uses 3D printers and other tools to create costumes and special effects, had created plans for such a prosthetic, and had posted the plans for free. With an investment of $2,000 in a 3D printer (the cost of which has since dropped precipitously) and $5 or $10 in materials, McCarthy was able to print a device that looked like something straight out of a sci-fi novel. (A conventional prosthesis might have cost many thousands of dollars.)

This sort of application is of course incredibly, almost indescribably, powerful. For the first time, Leon could pick up a cup or a pencil with his left hand. He can even draw with that hand now. The prosthesis makes him feel, Leon said in a CBS video news report, "special instead of different." And the smile on his face as he said it could light up a room, or a world.

Leon is not the only beneficiary of the new technology. This is happening more and more often. Recently a three-year-old in Maui received a 3D-printed hand. It cost a mere $50 to build, compared to up to $40,000 for a commercially created prosthetic. (In fact, a nonprofit group called E-Nabling the Future built and donated the hand, so in this case the boy's parents paid absolutely nothing.)

The benefits of this sort of assistive technology are not limited to humans, of course. An Indiana Chihuahua named TurboRoo, born without front legs, now scoots around quite comfortably in a custom 3D-printed cart. Assorted fowl of various types now sport 3D-printed beaks, and 3D designer Terence Loring built a prosthetic leg for a duck named Dudley. (Poor Dudley apparently lost his leg in a fight with a particularly vicious chicken. Really. This was one tough chicken.)

Exoskeletons

In 2012, researchers at the Nemours / Alfred I. duPont Hospital for Children created a lightweight 3D-printed exoskeleton for Emma, a two-year-old child born with arthrogryposis multiplex congenita (AMC), a disease that causes curved joints and muscle weakness. Ordinarily unable to move her arms, with the suit (called a WREX, for Wilmington Robotic Exoskeleton), she can move about and play. Since the version of WREX that Emma uses is made with a 3D printer, any broken or worn parts can be printed and replaced in hours, rather than the days or weeks it might have taken to repair or build a machined device. And when she outgrows the "jacket" to which the exoskeleton's components are attached, the team simply prints her a new, larger one.

Bone and joint surgery is another area in which 3D printing is having an impact. For instance, the Mayo Clinic began using 3D-printed custom hips in order to reduce the failure rates they'd been experiencing with metal-on-metal replacements. More recently, doctors at the University of Southampton completed a hip replacement surgery in which a titanium implant was 3D-printed after first using a CT scan to determine the exact measurements required.

But the 3D printing revolution's greatest health-care-related achievement so far may be robotic exoskeletons, like the one built by 3D Systems and Ekso Bionics in early 2014. Amanda Boxel, a patient who had become a paraplegic in a 1992 skiing accident, was fitted with a custom-fitted exoskeleton precisely printed from exacting scans of her spine, shins, and thighs. The result was, in effect, a robotic suit that allowed Boxel to walk. (Even before that, Boxel had successfully tested and used eLEGS, an exoskeleton created by the same company, then known as Berkeley Bionics. The difference is that the 3D Systems / Ekso Bionics device is a "bespoke" suit,

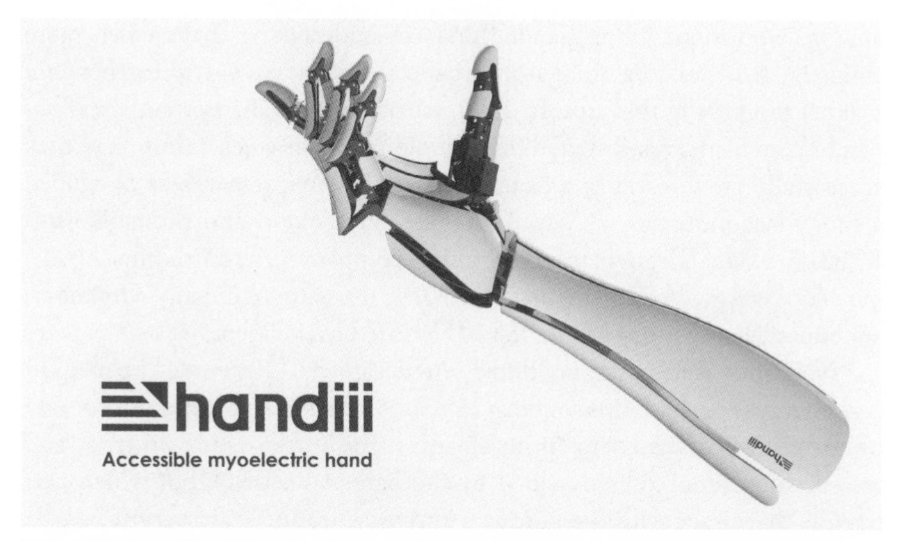

This is the handiii, a 3D-printed, smartphone-controlled prosthetic hand and forearm from exiii. Unlike conventional prosthetics costing many thousands of dollars, the goal is to sell this one for around $400. The project was runner-up for the 2013 International James Dyson Award. *Photo courtesy of exiii.*

a custom-3D-printed exoskeleton based on exact measurements taken from scans of Boxel's body.) In both cases, though, the result was that a woman who had been confined to a wheelchair was now ambulatory. Amanda can stand. She can walk. She can—as she said in a recent interview—look people in the eye instead of having to look up at them. That's simply an astounding feat; it's as if a Philip K. Dick or Isaac Asimov novel had come to life.

Bioprinting

As if robotic exoskeletons weren't impressive enough, consider the 3D printing subgenre known as *bioprinting*. While engineering students recently printed a plastic artificial kidney—itself an impressive development that could save countless lives once it becomes workable—plastic itself may become passé. It is becoming increasingly feasible to "bioprint" actual cellular material, and that could mean everything from a patch of skin to an entire organ.

In bioprinting, a device that looks (and acts) something like a standard 3D printer is used to layer gel-suspended cells, forming them into a shape appropriate for the job at hand. This is an undertaking that's much more complex than building things out of metal or plastic, but researchers are making progress in this area. In 1999, scientists printed a synthetic scaffold of a human bladder and coated the scaffold with living cells; from that, they successfully grew working organs. A few years later, researchers developed a functional miniature kidney that could filter blood and produce urine. In 2010, a San Diego–based bioprinting company printed the first working blood vessel. And in January of 2014, the same company, Organovo, announced the delivery of the first 3D-printed liver tissue.

Now, sometimes using modified or redesigned 3D printers, the medical community—or start-ups looking to sell to the medical community—are using printers designed to "print" living tissue (often made up of a "gel" grown from donor cells provided by the patient himself), but which also provide that tissue with the nutrients and oxygen it needs to survive.

It sounds incredibly complex, and it is. Nonetheless, there exists a growing DIY bioprinting community dedicated to building bioprinters (often hacked versions of old printers found lying about) to be used in a field that's becoming known as *regenerative medicine*. (One particularly well-written online set of DIY bioprinter instructions has garnered about 250,000 views.)

As with any other technology, there's no way to avoid controversy, especially when that technology is used to treat—or in some cases, literally rebuild—people. The whole health-care-related 3D-printing scene sounds vaguely Frankensteinian, of course, and there are those who would argue that we are treading on (literally) sacred ground. This may be especially true once people realize that some research could involve merging human cells with nonhuman cells. As one Gartner researcher noted somewhat understatedly in a 2014 *International Business Times* interview, "We think that once folks begin to realize, depending on their backgrounds, there will be some who will be concerned about this possibility."

It is a creepy thought, and while there may well be perfectly acceptable medical reasons for merging human and nonhuman cells, it raises troubling questions: Could we create a "chimera"—an entity that was part human and part animal? And if we did, why would we do so? What purpose would it serve? And if a new species were thereby created, would it be considered human? What rights should it have? Thorny issues for sure. (Of course, people who object to this sort of research might also be troubled by the fact that we're already using animal heart valves to replace faulty human ones, and have been doing so for many years.) In any such research, there is a moral line to be drawn at some point, and it's difficult indeed to determine in this case exactly where that line should be.

On the downside . . .

Printing Weapons

In addition to the vexing moral issues and legal concerns already noted, the 3D-printing "elephant in the room" is the fact that people have begun using these devices to print fairly sophisticated weapons. And, being a digital commodity, the files that describe *how* to create those weapons are now all over the Internet. (In fact, you don't even have to print the entire weapon. The lower portion of the AR-15 semiautomatic rifle, for instance, is the piece that's controlled, serialized, etc. And people have been successfully printing lowers for a few years now.)

The genie really escaped (or was unceremoniously shoved out of) the bottle in the spring of 2013. That's when Defense Distributed (run by an anarchistic former law student named Cody Wilson) demonstrated the world's first fully 3D-printed gun, the "Liberator." Consisting of sixteen parts, fifteen of which are made of plastic (the firing pin is a common metal nail), the gun was printed, says Wilson, to demonstrate that governments cannot control the creation and spread of weapons. Defense Distributed used a leased secondhand Stratasys Dimension SST printer to create the gun. (That printer is now a bit out-of-date, but you can buy one on eBay for around $6,000.) Wilson was forced to take the plans offline, but we're talking about digital items, here; by the time Wilson complied with that order, the file had already been downloaded 100,000 times.

Some months later, Valencia, California, engineering firm Solid Concepts produced the world's first 3D-printed *metal* gun, firing more than fifty rounds from it with no problems. The weapon is an exquisitely machined copy of the venerable 1911 .45 caliber semiautomatic that

A company spokesperson holds the Solid Concepts 3D-printed metal gun. *Photo courtesy of Solid Concepts.*

was carried by the US Armed Forces for years, and which is still a popular pistol and a very effective weapon. (The company later submitted the weapon to a torture test, firing five hundred rounds through it. Again, the gun showed no signs of being overly stressed.)

Solid Concepts' 1911 handgun was not cheap to create, having been made of 3D-printed stainless-steel parts created on a fairly expensive industrial-grade printer. But prices of those printers are falling; sooner or later, such devices will be within reach of any serious hobbyist. (In any case, many criminals are frighteningly well funded: Drug cartels or larger street gangs can easily afford to purchase such a machine; when they do, they will have the means to create an almost inexhaustible supply of inexpensive, easily repairable, completely untraceable weapons.)

These are issues that are more than simply worrisome; they're downright scary. (*Wired* included Cody Wilson in a list of "The Fifteen Most Dangerous People in the World," citing the fact that he had "created the first platform devoted to sharing the blueprints online for free to anyone who wants [a gun], anywhere in the world, at any time.") This is serious stuff, and there are no easy solutions. People *do* have certain rights, after all, and it would be difficult to legislate away this sort of problem without trampling on several of them. Then again, people generally do *not* have the right to endanger others—and 3D-printed weaponry would seem to provide a simple, affordable, and untraceable way to do exactly that, if that's what one had in mind.

Keeping Things Legal

The intersection of 3D-printing technology and the law brings us to an interesting juncture at which, as Washington, DC, attorney Michael Weinberg has pointed out, some "novel legal issues" will arise. In this case, that's a polite way of saying that there are going to be raised some very important questions to which we simply do not have any good answers.

Crucially, *copyright* law generally does not apply to 3D-printing issues (unless the article in question is actually art), but *patent* law may. As Weinberg notes, "While most modern songs, movies, and pictures are protected by copyright, the same cannot be said for physical objects."

A New York attorney with whom we spoke said that analysis of the effects of patent and copyright law on 3D-printed items inevitably turns into conjecture. After all, it's early in the evolution of the law as it applies here and, as he noted, "Until some of this stuff gets litigated, it's all just guesses."

However, the attorney (whom, for the sake of convenience, we'll call Mr. Finch) noted that some aspects of the law do at least seem to lend themselves to some preliminary scrutiny. For the sake of our analysis, we'll consider the creation of something fairly simple: a hinge.

So, let's set the scene. There you are, patting yourself on the back for having successfully refinished those old kitchen cabinets. As you're preening and posturing and considering rewarding yourself with a glass of your favorite beverage, you glance up and notice that the job you thought you'd just completed in fact remains undone: One of the hinges on a cabinet door is broken. Sadly, you examine it, trying to see if there's any way to repair it. There isn't, but then it dawns on you: You have a scanner, a 3D printer, and dozens of perfectly good hinges to serve as examples. You're home-free. All you have to do is scan one of the good hinges and print a replacement. Easy. Once again, you're feeling self-congratulatory. But think for a moment: When you scan and print that new hinge, will you be breaking the law? Will the hinge police show up at your door, along with a posse of goons from the Acme Hinge Company? You can relax. The answers, respectively, are most likely *Well, probably yes*, and *Probably not*. Here's why.

The process of 3D printing begins with the creation of a CAD/CAM (computer-aided design or computer-aided manufacturing) file of one sort or another. The file could be created by freehand drawing, in which case, at the least, the creation of the file itself is not legally problematic. But what if the file were created by *scanning* an object?

"Patents prohibit the making, using, or selling of patented items," says Finch. "It's unlikely that creating a 3D scan of a patented hinge would be considered 'making' a hinge, and, as a result, patent law likely would not prohibit scanning. Copyright law, however, does prohibit unauthorized *copying*, so if the item scanned is protected by copyright, scanning the item could be copyright infringement." Generally speaking, physical items are *not* protected by copyright, but if the item itself were a statue or a piece of jewelry (or a useful piece that nonetheless also includes obvious artistic elements), copyright could—and probably does—apply.

But what if, after having slogged through the legal morass of creating or acquiring a 3D file, we then *printed* the item? Ah, now things get touchy.

Finch notes, "It seems pretty clear that printing a patented hinge would be considered patent infringement. That said, there are a number of practical considerations that may prevent a patent holder from bringing suit. First, unlike copyright lawsuits involving music or movies, it's not enough to show that a user *downloaded* a CAD/CAM file; the patent holder also has to show that the user actually *printed* the file. Second, damages for patent infringement are 'reasonable royalties.' These damages add up when you're talking about a mass-produced product that violates a patent, but if you're talking about a one-off violation by a home user, you're likely talking about a few dollars or, in some cases, maybe even pennies. The recovery just isn't going to justify bringing a patent infringement suit."

So the bottom line is that producing a 3D-printed version of a patented object may in fact be illegal, but it's normally not worth the patent holder's time to prosecute.

But don't count on that being the case forever.

"Although I'd expect most infringement actions to involve copyright as opposed to patent, 3D printing certainly has the potential to expose home users to patent-infringement risks that they previously haven't had to worry about," says Finch. "Ultimately, we'll see adjustments made to patent and copyright statutes as well, although it's hard to say what form those changes will take."

I think the interview is over, but then he adds, "Of course, the lawyers are going to be the big winners in all of it." That might have been accompanied by a little lawyer-ish smile, but I'm not sure; after all, how can you tell when a barracuda smiles?

But Back to Those Houses

We started off this chapter by talking about houses, so now let's see what housing and 3D printing have in common.

Obviously, building a home is a complex, expensive undertaking. A home is generally the costliest purchase most of us will ever make, and it serves as both shelter and investment. We're comfortable with the idea that homes cost a good deal of money. Why shouldn't they? They require dozens (or hundreds) of skilled workers, extensive planning, an enormous supply of materials, and many months to create.

But construction is yet another industry that 3D printing may be poised to transform.

Consider the work of Dr. Behrokh Khoshnevis and his students at USC. Working in the university's Viterbi School of Engineering, Dr. Khoshnevis has come up with what he calls "contour crafting." Essentially, this boils down to the robotic construction of buildings using, you'll not be surprised to hear, what amounts to a huge 3D printer that extrudes not plastic or metal, but concrete. As with other forms of 3D printing, contour crafting involves spraying a layer—this time of concrete—that is automatically shaped and smoothed and allowed to dry. Since Khoshnevis's machine is printing walls, the "printer" is quite large: actually a gantry

USC's Dr. Behrokh Khoshnevis is the inventor of contour crafting, which can be used to quickly "print" homes and buildings. *Photo courtesy of Dr. Behrokh Khoshnevis.*

with a "printhead" that is really a canister of concrete whose motion on the gantry is controlled by a computer.

Using the new process, Dr. Khoshnevis and his team have succeeded in creating six-foot walls made of six-inch layers of concrete extruded with a special hardener so that they'll keep their shape while drying. Khoshnevis believes that using the process, one could print an entire 2,000-square-foot house in less than twenty hours. (He's even suggested that by placing the apparatus on rails, one could quickly print an entire neighborhood. It may sound unrealistic, but keep in mind that Chinese experimenters using large-scale 3D-printing techniques to create individual walls printed ten houses in less than twenty-four hours.)

The possible ramifications of this go far beyond simply being able to create suburbanites' homes for less money in less time. There are ways in which 3D printing on this scale could improve our world—and other worlds.

Khoshnevis has experimented with techniques that could one day help astronauts build spaceports and shelters on Mars. Since there is no water available on Mars to make concrete, the USC team experimented with a form of laser sintering (which uses heat and pressure to form a mass), the idea being that astronauts could use dust as building materials.

Here on Earth, the process could save lives, given that construction workers are engaged in what may be the world's most hazardous occupation,

This is a computerized rendering of contour crafting in progress. Note the gantry surrounding the structure. Also note the curved walls, which are generally stronger than straight walls. *Image courtesy of Dr. Behrokh Khoshnevis.*

with 400,000 people injured every year, and between 6,000 and 10,000 killed on the job.

And finally, consider the fact that this sort of technology could allow us to quickly and easily create reasonably priced housing in impoverished areas here and around the world. While many of us wish for bigger or nicer homes, there are millions around the globe who have *no* home whatsoever. This is one way (surely there are others) to help provide those people with decent shelter at a price we can afford.

Now Available to All . . . Well, Some.

The eLEGS and Ekso Bionics exoskeletons used by Amanda Boxel grew, as did so many of our new technologies, out of military research. A few years before building eLEGS, Berkeley Bionics designed HULC, or the Human Universal Load Carrier. Sponsored by Lockheed Martin, HULC is an untethered, hydraulic-powered titanium exoskeleton designed to allow a soldier to carry up to two hundred pounds of material, thus helping to prevent injuries and reduce fatigue. In a sense, HULC—and the civilian exoskeletons that derive from it—are what some researchers like to call "wearable robots." The medical applications for paraplegics, stroke victims, and amputees are immediately obvious, and we can't be sure when we'll exhaust the possible uses of such devices—if such a thing is even possible. As Ekso Bionics CEO Nathan Hardy has commented, "There's a huge wave of human augmentation coming. [The field is] in its infancy."

It's coming, but it won't be cheap; at least, not at first.

Once on the market, Indego, a twenty-six-pound exoskeleton from Cleveland, Ohio–based Parker Hannifin, will probably cost in the neighborhood of $67,000—or about as much as a brand-new all-wheel-drive BMW 535i. ReWalk Robotics' ReWalk exoskeleton, marketed by Israeli company Argo Medical Technologies, is said to cost even more, about $85,000. And the latest Ekso Bionics suit is pricier still; it sells for about $100,000, and is presently used mainly by rehab clinicians for training purposes. (At press time, of all three devices, only the ReWalk system has been cleared by the FDA for home use.)

So, these devices are not what one might call affordable. Eventually, though, insurance companies may pick up part of the tab, since their use would help to avoid a host of ailments that generally result from sitting for long periods in a wheelchair, including pressure sores, diabetes, and bone loss. Thus, there's a reasonable argument that paying for such devices could ultimately *save* the companies money.

Still, what we're seeing is typical of the product development and adoption curve for technology: At first, devices are available only to researchers

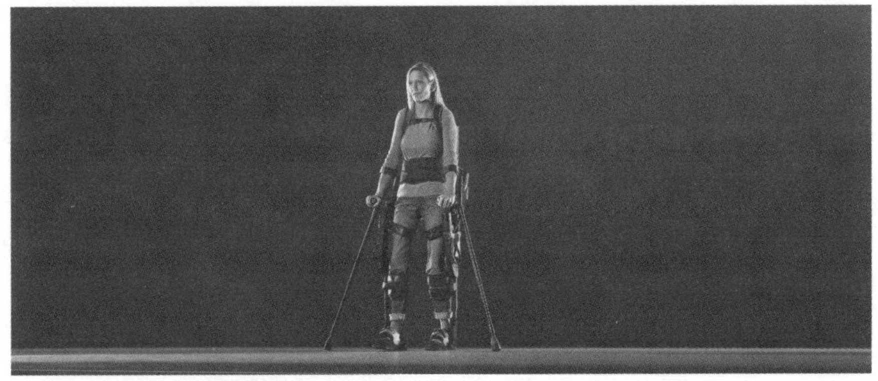

An Ekso Bionics "Test Pilot" and "Ambassador" uses the company's powered exoskeleton. *Image courtesy of Ekso Bionics.*

or well-funded government or corporate entities. Eventually the price drops and barriers are removed to the point that more people can avail themselves of the technology. Finally, the technology becomes ubiquitous, priced such that a large percentage of the population can acquire and use it.

Which of course is exactly what we've seen with 3D printers: The pricing/availability arc is now almost to the point that, as Dr. Bowyer envisions, just about anyone who wants a simple 3D printer can manage it. (And of course, the prices of the more-sophisticated printers, such as the ones used by Boeing and other industrial behemoths, will also continue to drop until those too are affordable to a much larger portion of the population than is the case today.)

The exoskeletons currently being tested will probably never be *cheap*; they do, after all, require not just 3D-printing technology, but also electrics, electronics, hydraulics, processing power, software, training, and so on. But they will certainly become much more affordable than they are today. (And the fact that they exist in this form at all is already quite an improvement over early attempts that cost millions of dollars and were, like General Electric's 1960s experimental 1,500-pound behemoth, in the end, unusable.)

Like the automobile, they will almost certainly go from being items available only to the very rich to being relatively common tools accessible to many who need them. This would seem inevitable, especially considering that the industrial and military uses of such tools, along with the obvious medical applications, will no doubt help to encourage continued research, refinement, and competition.

Like many technologies, 3D printing remained for years in the hands of wealthy and powerful people, governments, and larger corporations. Really, that's only fair; at least at first, these were the people who helped to invent the technology, and who for a very long time used it to create complex, sophisticated items that held little mass-market appeal. (Seriously, do you really want a ninety-foot drone? Well, okay, you might *want* one, but would it be wise to allow people to build such things? Even if you wanted to build one, would you want your *neighbor* to be able to build one?)

Now, though, the technology has been improved and honed and fine-tuned until it is possible to use small versions of these industrial machines to create useful household items. (If you're not sure that the prices of 3D printers will ever come down to mass-market levels, consider the Peachy

3D printer. Canadian company Rinnovated Design turned to Kickstarter to fund the creation of Peachy, which costs a stunning $100. Not surprisingly, the company's $50,000 target was successfully funded—to the tune of over $651,000 Canadian, about 1300 percent of the company's original goal.)

So perhaps Adrian Bowyer's vision will come to pass. It seems quite likely that, one way or another, 3D printers are destined to become ubiquitous tools that all of us could have in our garages, offices, or dens someday. It's difficult to imagine the explosion of innovation that could occur when that happens. Perhaps the changes will be as enormous and as transformational as those wrought by the introduction of the computer itself.

In fact, that's exactly how Dr. Bowyer sees it: "The personal computer made everyone a processor of information," he says, "and look how that turned out. RepRap will make everyone a manufacturer; let's see how *that* turns out."

Of course, one of the things you might want to print for yourself is a nice new case—in any of a variety of colors—for that fancy little computer you're carrying around in your pocket: your smartphone. Who knows? Someday you may be able to print the entire phone.

We'll tackle the smartphone revolution next.

CHAPTER NINE

Don't Call Me, I'll Call You: The Rise of the Smartphone

Terrified of being alone, yet afraid of intimacy, we experience widespread feelings of emptiness, of disconnection, of the unreality of self. And here the computer, a companion without emotional demands, offers a compromise. You can be a loner, but never alone. You can interact, but need never feel vulnerable to another person.

—SHERRY TURKLE, *THE SECOND SELF*

BUT, WHERE WILL SUPERMAN CHANGE INTO HIS COSTUME?

ONE DAY A YEAR OR SO AGO, I WAS DRIVING DOWN A MAIN STREET IN LINCOLN, Nebraska (yes, there are main streets in Lincoln, Nebraska), and out of the corner of my eye, I saw something that seemed somehow . . . *out of place.* For a moment, I couldn't quite put my finger on it; I just knew that I'd driven past a sight that seemed wrong. *Foreign*, somehow. It rattled me, because I couldn't quite reconcile what I was seeing with what seemed "normal" for life in 2014 in Lincoln. I had to think about it for a few moments, before I realized what I'd seen and why it seemed off to me.

It was a pay phone. Specifically, it was a woman at one of those half-sized, unenclosed pay-phone booths, smoking a cigarette while standing out on the sidewalk as she spoke into the handset. I hadn't seen a working pay phone in ages! It was such an unusual occurrence that, for a moment, my brain refused to credit what it had seen.

In this age of the ubiquitous cell phone, there simply aren't very many pay phones left. In 2002, there were still over two million of them in service. Ten years later, that number had shrunk to about 250,000. (An industry group says that some 1.7 billion calls per week are still processed by pay phones, "so clearly they are valued by many people in this country." But then, that's from what amounts to a pay-phone lobbyists' group. There is some evidence, though, that for those with limited cellular access, the remaining pay phones do provide a sort of lifeline.)

THE FIRST CELLULAR PHONES

But let's face it, pay phones are on their way out, and they have been since Motorola introduced the cellular phone in 1983. That first phone, the Motorola DynaTAC (invented by a Motorola team led by engineer Martin Cooper), sounded the death knell for pay phones, but it didn't matter too much at the time, because you couldn't have afforded one: That clunky phone (it stood about thirteen inches tall and weighed 1.75 pounds) sold for about $4,000 at a time when the average yearly salary was about $15,000. It was a technology reserved, as are many technologies at first, for the wealthy, the powerful, and the (in this case, quite literally) well-connected. (The original DynaTac was succeeded in 1989 by Motorola's MicroTAC 9800X, which cost a hefty $2,400 to $3,400, and which ushered in an era of increasingly smaller and less-expensive "flip phones." These featured mouthpieces that swung up against the body of the phone, reducing its overall length, but then flipped down and out to provide a microphone into which one spoke.)

As with many other technologies, this one didn't stay unaffordable for long.

Early on, revenues in the nascent wireless industry were bolstered by sales of bag phones and car phones that were in some ways more convenient and more effective than the early portable models, but the portables kept getting smaller, lighter, and less expensive. Along the way, manufacturers began integrating more tools: voice mail, fax, paging, contact lists. Soon they were called "feature phones." No longer tethered to a car (or packed in a bag), they were now small enough to carry in a pocket, and convenient and powerful enough to have become useful business appliances. (They eventually became so small that consumers began complaining that the user interface was unacceptable; manufacturers had designed themselves into a

In 2007, 34 years after he and his team invented it, engineer Martin Cooper poses with a prototype of the first commercially available cell phone, the Motorola DynaTAC 8000X. *Image licensed under the Creative Commons Attribution-Share Alike 3.0 Unported license.*

bit of a hole: Their products functioned perfectly well, but they were so tiny that consumers perceived them as difficult to use.)

The Rise of the Smartphone

By the 1990s, manufacturers were starting to look for the Next Big Thing in mobile telephony. IBM found it first, but, ironically, the company may have found it *too* soon: The rest of the world was not yet ready for Simon, arguably the world's first smartphone. Released by IBM in 1994, only fifty thousand units were sold, and the product was quietly shelved.

Though not a commercial success, Simon was recognizably the forerunner of today's iOS, Windows, and Android devices. At $800 with a service plan and a battery life of only about one hour, it was expensive, finicky, heavy (weighing about one pound), and about the size of half a house brick. Nonetheless, it offered a touch-sensitive screen (mono, of course), and the ability to create notes, send e-mails and faxes, and look up contacts. It was essentially a (slightly uncomfortable) marriage of the PDA (personal digital assistant) and a cell phone. (You can still find working models on eBay for around $200, but you cannot actually make telephone calls with them, because the analog cellular service used by the device disappeared long ago.)

IBM's Simon, released in 1994, was recognizably the first smartphone. *Image courtesy of user Tcomotcom, released into the public domain via Wikimedia Commons.*

Simon receded into the background, while ever more powerful true PDAs (the original Palm Pilot among them) took center stage.

But then came BlackBerry (originally Research in Motion, or RIM), which tested the communicative waters with the 1996 release of its RIM-900 pager, complete with keyboard. Then, three years later, RIM released the BlackBerry 850, followed in 2002 by the BlackBerry 5810, which for the first time featured voice-calling capabilities.

Palm—by then owned by 3Com—had come oh-so-close to grabbing the wireless brass ring in 1999 when it had released the Palm VII, which sported an antenna that could be used to connect not to a wireless telephony service but to the proprietary Palm.net network. Close, but no cigar. However, it was about to hit it big with its Palm Handspring Treo, followed by the PalmOne Treo 600, a multiband smartphone available in both CDMA and GSM versions, which ran Palm OS 5.2.1 and featured a 160 × 160 color touchscreen. (In the meantime, Sharp had in 2000 released the first camera phone, the Sharp J-SH04. At the time, many of us said, "Why would anyone want a camera on their phone?!" Little did we know.)

Now we had true smartphones: workable, useful, and more or less affordable marriages of the best of then-current cell technology and the handy PDA.

But the next giant leap in smartphone evolution—some might say it was something of a *revolution*—was about to occur. On January 9, 2007, fifty-one-year-old Steve Jobs (who, unbeknownst to us all, had only another five years to live), clad in his trademark black turtleneck and blue jeans, stood on a stage at the Moscone Center in San Francisco and announced the Apple iPhone. "Today, Apple is going to reinvent the phone," he said, and he was more or less right.

It let users employ a touchscreen to browse full web pages and to geolocate using Google Maps, and it provided access to the same iTunes music library and store that so many had become so used to on their iPods—only now you could use a touch-sensitive screen to *scroll* through your music, your contacts, and other such items. Interestingly, that first introduction, though it mentioned several apps that came with the phone (photo management software and a calendar, for instance), made no mention of Apple's App Store, which would quickly become a retail behemoth, selling literally billions of apps and driving sales of the phones (and later, tablets) themselves.

The original Apple iPhone (left) and the second-generation phone, the 3G (right). *Image courtesy of Dan Taylor, licensed under the Creative Commons Attribution 2.0 Generic License.*

(Third-party apps would be added the following year, with the release of the next version of the iPhone.)

Some would say that it was Apple that really *made* the smartphone business, generating sales and interest in smartphones (Apple's and others') to such an extent that the smartphone business became the fastest to reach the trillion-dollar mark in human history.

But Are They Necessary?

There's more—much more—to the story, of course: new versions of the iPhone, the introduction of tablets, the wild popularity of phones and tablets powered by various versions of the Android operating system, and of course, Microsoft's ongoing (and some might say ill-fated) foray into smartphone and tablet territory. The phones continue to get more powerful, more full-featured, and—at least in terms of the relative power on tap—less expensive all the time. (Also, they now seem to be getting larger, rather than smaller. Currently, the "phablet," halfway between a comfortable phone and a somewhat smallish tablet, seems to be all the rage. Perhaps manufacturers are trying to come up with the absolute *worst* design possible, eventually arriving at a device that's too small to be a useful tablet and too large to be a comfortable phone.)

But, as impressive as these devices are, do we really *need* them?

"Modern-day smartphones—the Apple iPhone in particular—changed everything that consumers expect from their phones," says Neil Mawston, executive director at technology research firm Strategy Analytics. "The app market has transformed the phone into a virtual toolbox with a solution for almost every need."

Considering that your smartphone is actually a fairly powerful handheld computer (much faster and more powerful than the computers that sent men to the moon and back, as many have pointed out), there's not much they can't do in terms of helping out at work or play. Almost regardless of the situation, there's an app (or information accessed via your device's browser—the browser itself being an app, of course) that will locate you, help you, inform you, or entertain you.

In Vino Veritas

Take wine, for instance. It's . . . well, *complicated.* There's so much to know, and there's so much about it that most of us *don't* know.

Andrew Browne knows, though. Andrew is a certified sommelier who works in the San Francisco Bay area. He knows a lot about wine, writes magazine columns about it. Andrew lives and breathes wine.

Andrew also has a smartphone (well, of course he does), and on his smartphone is an app called Delectable, which he uses to find out about wine, log info about wine, and—not at all surprisingly—share info about wine.

"Delectable is probably the best-known wine app out there," says Browne. "The most useful thing about it is that it lets you find a wine instantly and tells you where you can get it—or better yet, you can set up your profile with a credit card and address and they'll ship it to you!" (Well, someone has to be making some money off of this, right?)

Sommelier Andrew Browne works in the San Francisco Bay area and utilizes various technology tools in his job, including the wine app Delectable. *Image courtesy of Andrew Browne.*

Wine aficionados are nothing if not social, so Andrew says that he likes the fact that Delectable lets you share (and receive) notes and reviews from others, and also check out wine scores from top critics and magazines.

"You can also tag your friends on Facebook, as well as find reviews and notes on wines you haven't tried. It's a very intuitive and interactive wine app."

As a professional "somm," Andrew has discovered that the social aspect of Delectable (and of similar apps) tends to be a very useful way to advertise oneself.

Delectable is among the most popular wine-related smartphone apps. *Photo by Daniel Dent. Image courtesy of Delectable.*

"I have a number of people who follow the wines that I taste and recommend on Facebook," says Browne. "People have followings for all types of things. For me it's wine."

Andrew's mobile device has become much more than a phone. In fact, the idea that it can make telephone calls has become somewhat incidental; in Andrew's case (and in the case of many others as well), he's carrying around a pocket communicator, a tool that allows for connectivity, entertainment, and information-gathering, but if he's like the rest of us (and he is, except for that whole *wine* thing), he only rarely uses his "phone" to make phone calls.

He's certainly not alone. According to a 2012 survey, smartphone owners were only infrequently using their devices to make actual phone calls; the phone app was the fifth-most-used app, trailing well behind the device's browser, social networking, music, and other functions. (At the time, surveys noted that users were spending over two hours per day staring at—but apparently not speaking into—their phones.)

More recently, the Pew Research Center has compiled data indicating that some 81 percent of smartphone owners send or receive text messages, 60 percent access the Internet, and 52 percent send or receive e-mail. In Britain, taxi app company Hailo surveyed users in 2014 and found that in the UK, the telephone app itself came in sixth place in terms of use, and some 40 percent of users said that they could do without the "phone" part of the smartphone altogether.

So, perhaps when Palm, RIM, and others released that series of phoneless PDAs, they were actually *ahead* of their time.

BACtrack: A Personal Blood Alcohol Measurement Tool

Of course, some tools are potentially more important than others, and since we're speaking of wine, consider what happens when one drinks a little *too* much of it.

Back in 2001, a University of Pennsylvania student named Keith Nothacker started wondering why, if one was not allowed to drive with a certain BAC (blood alcohol level), there was no good way for someone to determine his own BAC. Surely, he thought, there must be some readily available tool that would let someone know when he'd had one too many drinks and was better off staying where he was—or perhaps simply ensuring that someone else drove.

There wasn't. Nothacker saw both a problem and an opportunity. He started a company called BACtrack, with the goal of providing personal BAC monitors, to, as BACtrack spokesperson Stacey Sachs says, "help people make smarter decisions when consuming alcohol by offering the most accurate, convenient, and affordable breath alcohol testers."

It took years, but—aided by appearances on high-visibility television shows such *The Doctors*, *MythBusters*, and others—BACtrack took off. Nothacker no longer had to argue with Amazon buyers about whether this was a legitimate product category; BACtrack devices were selling well, providing a usable and affordable alternative to the clunky, $1,000, police-only devices that had been available previously.

The next step was to shrink the package even more, and to take advantage of the computing and communicative power of the smartphones everyone was already carrying around with them.

"In April 2013," says Sachs, "BACtrack launched the world's first smartphone breathalyzer, BACtrack Mobile. It uses Bluetooth to wirelessly connect to iOS and Android devices. BACtrack Mobile incorporates a police-grade alcohol fuel cell sensor and miniature internal breath pump to ensure consistent, accurate results. This was not just the first smartphone-connected breathalyzer at launch, but also the smallest full-featured (fuel cell technology-plus-pump) breathalyzer ever invented."

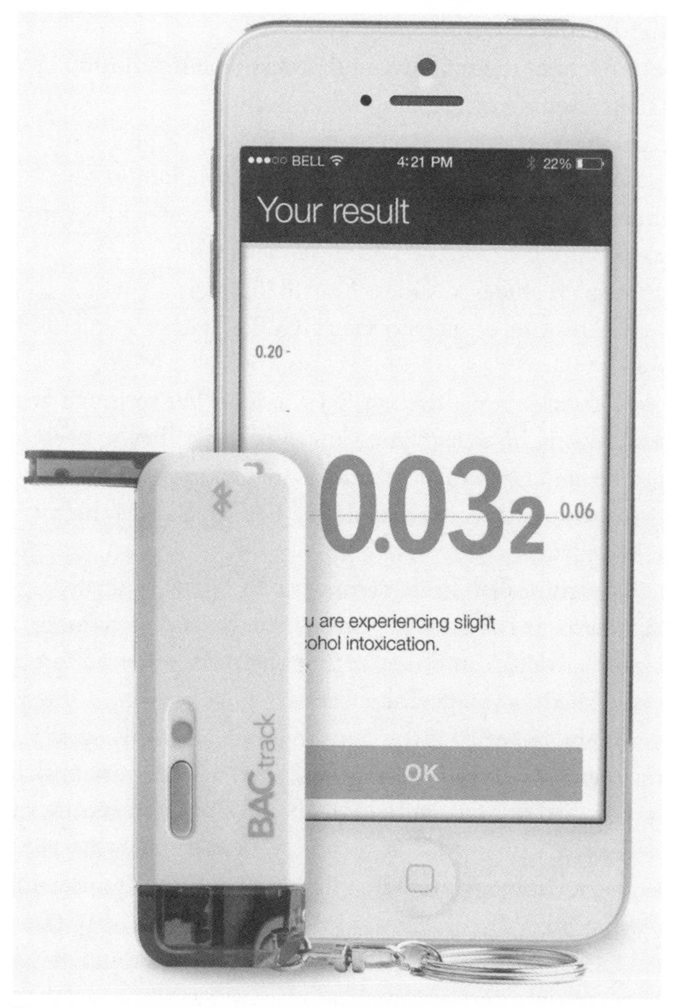

The ultra-portable BACtrack Vio is a keychain-size device that connects to your smartphone and can help you track your blood alcohol levels. *Image courtesy of BACtrack.*

While earlier breathalyzers simply displayed a BAC result, BACtrack Mobile displays safety information at each BAC point, helping to educate users about how alcohol affects their bodies.

The computational power and connectivity of the smartphone allow a tool such as BACtrack to provide more than just a simple status readout. It can show a user approximately how long alcohol will remain in his bloodstream after he stops consuming alcohol.

"It's eye-opening to see how you might still be intoxicated in the morning after a late evening of drinking," says Sachs.

Additionally, BACtrack Mobile can save and display a user's drinking history. In theory, this helps the user to become much more aware of alcohol consumption habits, whether they are healthy or unhealthy, through visual BAC tracking.

The app also includes a "Guess Your BAC" feature, which lets a person guess prior to testing. A person can view her guesses and actual results over time.

It's more than a game, though, says Sachs. "We've found that breathalyzer users become better at guessing over time; they're becoming more knowledgeable about how alcohol affects them."

It's not going to save the world, but BACtrack does help to level the playing field a bit.

"There are online BAC calculators," notes Sachs, "and the DMV sends out charts that allow a user to estimate based upon number of drinks consumed, time, weight, and gender, but there are other factors that come into play in a BAC estimate. These include the strength of the drink, how quickly it was consumed, if it was consumed on an empty stomach, etc. We hope the product levels the playing field in that everyone now has access to information about how alcohol affects the body and can use it to make smarter decisions."

Of course, technology giveth and technology taketh away. Like any of the tools we've been discussing, it's possible to misuse BACtrack, and it's also possible that information collected by the tool could fall into the wrong hands. When that happens, a serious privacy issue arises; be sure to see chapter 12 for more about that.

Your Smartphone May Change (or Even Save) a Life

Wine (and drinking too much of it) are important, of course. But smartphones are also beginning to be used in even *more* critical circumstances—sometimes, life-altering or even life*saving* circumstances.

Every year in the United States, thirty-eight children die because they're left in hot cars. That's an ugly statistic, and it's made even uglier by the fact that there's often absolutely no reason for it; more than half of the time, a harried parent or caregiver simply *forgets* that there's a child in the backseat (perhaps she's distracted or dealing with another child) and heads off to an appointment or shopping or work, leaving a toddler or baby trapped in a car that very quickly begins to overheat. (How quickly? According to the National Highway Traffic Safety Administration and San Francisco State University's Department of Geosciences, a car's inside temps can jump 20 degrees in only ten minutes. And even when the outside temperature is only in the 60s, under some conditions, the temperature inside the vehicle can reach as high as 110 degrees.)

Intel is working on a smartphone-connected solution, though. The company has produced the Intel Smart Clip, which attaches to your child's seat harness; if the harness is fastened and conditions in the vehicle become potentially dangerous, the device can warn you of the danger.

If you're seeking more examples of potentially lifesaving smartphone apps, consider the variety of "panic-button" tools that are now available to provide help or security should you happen to be caught in a situation that appears dangerous. Some of these apps will raise an earsplitting alarm; many will phone or text your exact location and send a "Please help!" message to a trusted friend or to the authorities. (Some panic-button apps will even record audio and display a fake home page to conceal the fact that the application is running.)

Finally, consider a piece of medical technology that could save a life—or a village.

A Philadelphia start-up called Biomeme has created a mini lab DNA analysis tool that attaches to an iPhone and uses the phone's computing power and camera to test people for diseases and to run other biological diagnostics. The Biomeme team essentially has taken a common diagnostic tool, the thermal cycler or PCR machine—which can cost many thousands of dollars—and made it both portable and less expensive.

The result is a piece of equipment that's similar to thermal cyclers found in labs around the world, only smartphone-sized. It connects to your iPhone and includes everything you need to prepare a DNA or RNA sample.

After, say, a urine sample has been prepped, it is inserted into the thermal cycler, which can identify the presence of a certain DNA or RNA strand for, say, the flu. If the flu is present in the sample, the thermal cycler will multiply the RNA strands, and they'll begin to glow; if the flu is not present, the device won't be able to multiply it, so the sample will not glow. The machine then measures the luminosity of the sample using the phone's camera (remember our question about Sharp's first camera phone?) and determine whether the flu was present.

Imagine what the presence of one (or 100 or 1,000) of these would mean in developing countries or in rural areas, far from sophisticated labs and medical facilities.

We can assume that Motorola engineer Martin Cooper never anticipated such things when he and his team came up with that first cellular phone. What we're seeing is a huge leap—perhaps a series of huge leaps—from smartphone-related tools that are merely interesting and useful to ones that could turn out to be life-changing.

Addiction? Disorder? Or are we overreacting?

Then again, some think we may have become *too* fond of our smartphones.

Psychologist Larry Rosen compared four recent studies of smartphone/user interactions and concluded, "Our smartphones make us anxious, and that anxiety then gets in the way of our performance and our relationships. Some call it FOMO—Fear of Missing Out—or nomophobia—fear of being out of mobile phone contact—or FOBO—Fear of Being Offline. Regardless of what you call it, this disorder is a manifestation of anxiety, plain and simple."

Rosen points out that if a heavy smartphone user is separated from his phone he gets measurably anxious, and that anxiety affects his ability to attend to others around him, as well as to his work. Rosen notes, "Studies show that people are switching from one task to another every three to five minutes . . . and the stimulus for those switches is most often an alert or notification from their omnipresent smartphone."

Rosen says that we are all facing "a very difficult task of disentangling ourselves from an affliction (that seems like far too strong a word, but sadly may be an accurate one) that is rapidly approaching an anxiety disorder."

He recommends that if you're not sure how often you check your phone, try using an app such as Checky or Moment. Rosen did, and was surprised to find that by noon the first day he had checked his phone forty-seven times!

It's big business. Really big.

According to Gartner analysts, Apple, Samsung, and the rest of the smartphone vendors sold over 1.2 billion units in 2014 (with Apple's sales beating out Samsung's for the first time since 2011). That's a lot of smartphones; in fact, it's about 28 percent more than were sold in 2013. So if you're thinking that the market has been saturated and that sales should be leveling off, well, that's not going to happen for a while, because other markets (huge markets, such as Latin America, Africa, and China) are opening up even as it seems that everyone you know already owns a smartphone—or even two. (And also because people keep buying newer, supposedly "better" phones every year or two, as noted below.)

Here's how big the smartphone business has become: According to some estimates, the sale of *used* smartphones (not the sale of new devices, but the sale of refurbished ones) will be worth about $14 billion by 2017. And according to the latest research from Strategy Analytics, global smartphone operating profit reached US $21 billion in Q4 2014. That's not per year, that's in one quarter. And it's not a revenue figure—it's *profit*.

In other words, some people (people who are not you, most likely) are making a lot of money manufacturing, programming, shipping, and servicing smartphones. And whole other groups of people (who are *also* not you) are making money selling phones and providing wireless service and the customer service—however optimistic the phrase "customer service" might sometimes appear to be—to subscribers.

And about those "free" or subsidized phones . . . According to Neil Mawston, they're not really free—and only barely subsidized.

"Operators still make profits on hardware and also data plans," says Mawston. "It's just that those profits are delayed for six to twenty-four months while the monthly subscription is paid back by the consumer."

Neil Mawston, executive director at technology research firm Strategy Analytics, follows the mobile communications industry. *Image courtesy of Neil Mawston, Strategy Analytics.*

So, basically, you *are* paying for the hardware, but slowly. Just quickly enough, in fact, to be ready for the next generation of phones to be released about the time your contract is up. Surely this is a coincidence of the most astounding and serendipitous sort.

More recently, the major service providers have mostly done away with the standard two-year contracts. Now we can pay for our phones up front or spread out the payments over some period of time; either way, we're less likely to be locked into a contract that's either impossible or expensive to break. Other than that freedom, the only real difference is that in those situations, we now *know* we're paying for the phone instead of believing (falsely) that the provider is subsidizing the cost.

The Shifting of Telco Jobs

As with the other technologies we've discussed, the introduction of the cell phone (and a bit later, the smartphone) surely resulted in a terrific disruption in jobs—in this case, at the old-line telcos. For more than one hundred years—first through "Ma Bell" (really just a nickname for AT&T), and then later, the landline providers that swarmed in after the US government carved up the Bell System in 1984 and opened up what had been a monopoly to the "Baby Bells"—landline service was king. (At its peak, AT&T employed about one million people.) Ultimately, even with hundreds of allied smaller carriers and the original "Baby Bells," landline (or what the companies call "fixed-line") service was the first choice in electronic communications. In fact, it was essentially the *only* choice.

But that disruption, as huge and impactful (and sometimes painful) as it was, almost certainly resulted in a net *gain* of jobs, rather than a loss.

"The global mobile industry is now far bigger by revenue and units than the global landline industry ever was, so it is reasonable to suggest there has been a net jobs gain worldwide from the transition to mobile telephony," says Mawston. "When I did my MBA," he continues, "the well-worn theory of 'creative destruction' indicated that a destroyed industry will always be replaced by a bigger and/or better one. I think that theory generally still holds true—but with the increased outsourcing of human jobs to computers, robots, and other automated technologies, it may reach a point where the net jobs gain from creative destruction sometimes no longer holds true. For example, the Internet has made newspaper publishing a much less human-intensive industry."

Of course, the supposed net gain is small comfort if you were one of the casualties and hadn't had a chance to retrain or move to another position.

Mawston's example of the newspaper industry as one that has been hard-hit by the electronics and mobile revolution is a useful one, and apt, but it may be incomplete. Newspapers are indeed failing, but *news* is still very much in demand, now delivered via desktop, tablet, and other devices. (And today collected, aggregated, and delivered the way you like it, on your schedule, and with a great deal of immediacy. The question is whether news can be collected, aggregated, and delivered *at a profit.*)

If you're in the business of selling newsprint, your days are indeed numbered. But if you're a journalist, an editor, a media specialist, an illustrator, or a publisher (or an IT person, service provider, cybersecurity expert, or website designer), you may yet survive the demise of the once-great print-media giants and find yourself working for the digital arm of a large publishing combine, or possibly for a small, upstart electronic journal of the sort that could never have existed—let alone competed—before the technology-driven media revolution. (Or you might even profitably disseminate information through your own blog. Several examples come to mind of entrepreneurial sorts who have managed to turn their passions into blogs, and their blogs into—often fairly comfortable—livings. For instance, see blogs by Amy Jackson [FashionJackson], Dulcie Wilcox and Sarah Welle [two tarts], Kristen Doyle [Dine & Dish], and former *Washington Post* reporter Brian Krebs [KrebsonSecurity], among many, many others.)

Amy Jackson's Fashion Jackson blog, website, and Instagram images have attracted thousands of followers eager to learn about and view the latest fashions and accessories. *Image courtesy of the Fashion Jackson blog. Photo by Karl Mayer.*

And don't feel bad for the landline companies. Most have actually fared well enough, according to Mawston.

"Most major landline operators worldwide have successfully transformed into hybrid landline-mobile operators," he says. "Verizon Wireless, NTT DOCOMO, France Telecom, etc., all had deep pockets to buy their way into mobile. Those landline operators that have struggled in mobile, such as BT UK, have diversified into semi-mobile services such as Wi-Fi for enterprises or homes."

Not everyone managed to adjust, of course: Nortel stands (well, *fell*, actually) as testimony to some of the destruction wrought by the wireless industry during the 1990s and 2000s. Founded in Canada in 1895, Nortel struggled to adapt and attempted to shift from wired to wireless tech. The struggle failed, and the company, which had once employed almost one hundred thousand people worldwide, filed for bankruptcy in 2009. The company finally went under and sold off its remaining assets in 2011.

On the downside . . .

It'll kill you. Or make you stupid. Or both.

Okay, not really. Though there are those who fear that the cell phone (now evolved into the smartphone) might in fact do both.

Although anecdotal evidence and your own common sense might seem to indicate that the use of smartphones is creating a generation of feeble-minded goobers, each with the attention span of a gnat and the intellectual capacity of a slightly overripe rutabaga, there is scientific evidence that says otherwise.

It does seem reasonable to believe that smartphones (and similar devices) are shortening attention spans. As one writer commented, "Screen-based activities can take upward of eleven hours of a teenager's day, and many demand rapid shifts of attention: quick camera cuts in videos, frenetically paced games, answering questions in multiple apps, not to mention web design that invites skimming. And we often do all this simultaneously, so attention bounces between two (or three or eight) fast-paced tasks. The theory is that the brain's plasticity turns this quick mental pivoting into a habit, rendering us unable to sustain attention."

Nonetheless, there's little actual evidence that attention spans are shrinking. And, although it's true that mental tasks can change our brains, the impact is usually modest. (Hence the recent furor over "brain training"-based programs that claim to make you—or your student—smarter and faster through exercises that "strengthen" the brain. One Texas company was recently forced by the FTC to stop making claims that its games could "permanently improve a child's focus, memory and school performance.")

So why does it seem that attention spans have contracted? As the same writer notes, "It may be that digital devices have not left us unable to pay attention, but have made us unwilling to do so."

Thus, it may be that we *can* still pay attention, but we choose not to, instead bouncing back and forth between multiple screens and multiple activities on multiple devices. (And, as any teacher or mid-level manager will be happy to point out, it doesn't much matter if a student or employee *can't* pay attention or simply *doesn't* pay attention; the result is likely to be the same: failure of one sort or another.)

So, your addiction to digital devices may not ruin your attention span. On the other hand, it actually *might* kill you, as it turns out.

For a number of reasons, the number of traffic fatalities in the United States has been falling steadily: Vehicles are safer, fewer people are drinking and driving, and most people now use their seat belts. And yet, *pedestrian* deaths are on the rise. In 2012, pedestrians accounted for 14 percent of US traffic fatalities, compared to 11 percent in 2007.

Why? Some are saying that the increase in pedestrian carnage is due to a phenomenon known as "distracted walking." In other words—yes, you guessed it—too many of us are apparently walking down the street (and through intersections) with our heads buried in our phones or tablets. Multiple studies seem to point to "distracted walking" as a major culprit in the rising numbers of pedestrian injuries and deaths. (Of course, not all traffic fatalities are dropping, and some of the ones that are not are in fact also connected to the use of mobile devices: More than 55 percent of drivers admit to using a mobile phone at least some of the time while driving. And "distracted driving," which includes texting, talking on one's phone, and other such activities, has become a leading cause of death of among teenagers.)

Texting while driving has become a leading cause of accidents among young people. This driver isn't looking at the road or at oncoming traffic. *Image courtesy of Lori Garris.*

As one *Washington Post* reporter wrote, "A 2012 study in the journal *Injury Prevention* found that nearly one-third of pedestrians at twenty high-risk intersections in Seattle were observed listening to music, texting, or using a cell phone. Those who texted . . . were four times more likely to display at least one 'unsafe crossing behavior,' such as ignoring traffic signals or failing to look both ways."

A year later, an Ohio State University study found that in a sampling of one hundred hospital emergency rooms, the number of injuries resulting from pedestrians using cell phones had more than doubled between 2005 and 2010, and that people between the ages of sixteen and twenty-five were the ones most likely to be hit while distracted.

So, your phone may not be making you stupid—it may just be making you *act* stupid. And the line between the two is very, very thin.

Changing the Game

As we've noted, phones these days are really "computer handsets," offering a great deal of computing power and the ability to run applications, surf the Internet, play games, and offer voice-guided directional (and other sorts of) assistance. And considering the cost of those first clunky phones, it's amazing how affordable they've become. (Even at $600—about the cost of some of the better handsets—today's smartphone today will run you about one-seventh of the cost of that first mobile device—and a good deal less than that in terms of adjusted dollars.)

It's difficult to overestimate the impact of the smartphone. As Mawston says, "Smartphones are personal assistants, digital books, pocket computers, mobile wallets, cameras, and entertainment portals all wrapped up in one handy device. Smartphones are becoming more, not less, important to daily life."

Richard Lieberman, in *Your Job and How Technology Will Change It*, talks about "general purpose technologies." These technologies, he says, are the game changers. "They represent unique technological breakthroughs that completely change the way people live and work. They include the internal combustion engine, electricity, and steam power. These general purpose technologies," says Lieberman, "result in innovation to business practices and processes, and enormous productivity gains for the worldwide economy."

That's what's happened with smartphones, and that's why they're an excellent example of a technology that has helped to level the playing field. Smartphones have proven to be game changers. They've definitely disrupted economies and upset workforces, but along the way they've also created completely new economies and offered countless opportunities—the last of which we've surely not yet seen—for the invention and delivery of new products, new services, and new ways to communicate.

In the beginning, these devices were reserved for the wealthy and powerful; now, for better or for worse (or for both, most likely), eleven-year-old schoolkids walk around with the latest Apple, Android, or Windows gizmo in their pockets, purses, and backpacks, and with them they socialize, communicate, and explore. They hold in the palms of their hands the power to entertain and to be entertained, to inform and be informed.

What they'll do with that power is anybody's guess, of course.

One thing those eleven-year-olds *will* do with their phones, of course, is socialize. They may not do it wisely, and they may not be aware of the potential dangers of social media, but almost all young people are using their phones and other devices to socialize with their peers. How they do so—and what the risks may be—are the subject of the next chapter.

Chapter Ten

We're Following You: Social Media

The Web is more a social creation than a technical one. I designed it for a social effect—to help people work together—and not as a technical toy. The ultimate goal of the Web is to support and improve our weblike existence in the world. We clump into families, associations, and companies. We develop trust across the miles and distrust around the corner.

—Sir Tim Berners-Lee

No News May *Not* Be Good News

It's eerily quiet this evening in Pyongyang, the capital of North Korea. No one is posting a thing. No tweets. No Instagram photos. No Foursquare check-ins. For the time being, the vast social network that hums and buzzes so incessantly around the rest of the world is silent here.

There's a reason for the digital silence. News reports are filtering back to the West that North Korean dictator Kim Jong-un may have been overthrown in what is being referred to as a "silent coup," one in which the young leader has not been ousted so much as simply ignored; he's now, some are saying, merely a figurehead, while the country's Organization and Guidance Department (the OGD, still loyal to the policies of his father) actually runs the country. (The rumors turned out to be overblown, of course. Several days later, things were back to what passes for normal in North Korea.)

At the same time, at the college bars around O Street in Lincoln, Nebraska (only a couple of blocks from Memorial Stadium, where the Huskers play), social networking—and probably several other forms of networking—are in full swing. In the past day or so, in an area of perhaps ten blocks, there have been a number of Foursquare check-ins, several Instagram posts (some of which come from the bar owners themselves: "$3 fireball shots! $3 tall boys!"), and dozens of tweets, some with photos, some without.

Almost seven thousand miles from North Korea, at the White House, perhaps the most visible symbol of political power in the United States, several people have checked in via Foursquare within the past few hours, even on a quiet Sunday night. A few have posted Instagram photos. A few dozen have tweeted from within a square block area surrounding 1600 Pennsylvania Avenue (including some who are enjoying the novelty of taking a "White House selfie" next to the iron fence surrounding the North Lawn).

So, as one might expect, a dictatorship faced with crisis tightens its iron grip around communicative technologies, and the ether goes silent. The White House, surely also crisis-ridden, but somewhat less likely to openly trample on the rights of its citizens, remains an alluring presence, drawing visitors young and old to gaze at the home of the president, to comment digitally about it, even to protest it—because in the United States, after all, we're supposed to be free to do that. (Just how true that really is, and how many rights actually get trampled on, and by whom, is the subject of another—perhaps more important—book.)

But the pertinent question as far as we're concerned is this: Sitting in my basement office in the Midwest, how am I able to know what's happening on social networks across the world, or even across town?

To answer that question, we have to take a few steps back—all the way back to the beginnings of social networking.

The Dawning

In the 1970s and 1980s we had online communities such as CompuServe, Prodigy, and America Online (AOL), but the closest thing we had to true social networking were bulletin board systems (BBSs) that were, at least at first, used mostly by academics and computer hobbyists.

In the early 1970s, there was e-mail ("electronic mail," in the somewhat more formal argot of its formative years), the first one of which is said to have been sent by an ARPANET electrical engineer and programmer, the late Ray Tomlinson, who directed that first e-mail to himself as a test. (It is to Ray, by the way, that we owe the use of "@" to separate a user's name from the name of the machine, or later, the name of a domain.) Ironically, that e-mail has not been preserved, and Ray assumed that it was some insignificant string of meaningless characters; it was a test, after all. (As Ray noted in a 2009 *New York Times* blog article, "The test messages were entirely forgettable, and I have, therefore, forgotten them.")

It wasn't until about twenty years later, on March 11, 1992, that Nathaniel Borenstein, a researcher and multimedia developer working for Bellcore, a small telephone company that would soon be acquired by SAIC (Science Applications International Corporation), sent what might be the first e-mail attachment: a photo of the company-sponsored barbershop quartet, to which he belonged.

In the late 1990s, things began to change, and very quickly.

By 1997, there were one million websites. (This may be about the time we began to lose count, actually; it's now difficult to tell how many sites

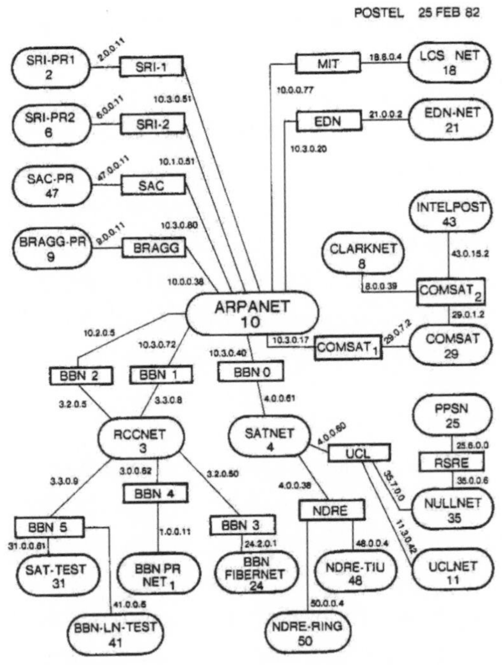

This was the entire Internet in 1982. The ovals are sites/networks (some sites included more than one physical network), while the rectangles are individual routers. *Image in the public domain.*

there are, partly because that may depend on how we define "website," and also because they appear and disappear with such fluidity. Some put the number of websites operating today at around thirty trillion, but that's just a wild guess.) The first blogs appeared, and AOL released Instant Messenger, which allowed members to chat.

Also in 1997, a company called SixDegrees.com somewhat presciently began allowing users to create profiles and list their friends. Users could send messages to other users, and could in fact invite friends to become members. (The site claimed well over three million members at its peak, and sold in 1999 for $125 million, after which it was essentially never heard from again.)

It's beginning to sound pretty social, isn't it?

Friendster and Myspace

In 2002, programmer Jonathan Abrams founded Friendster, a social networking site that soon had some two million users. By 2004, though, it had been overtaken in terms of popularity by Myspace (then called MySpace). (Things move fast in the world of technology, and social networking is no exception. These days, Friendster, now headquartered in Kuala Lumpur, Malaysia, has been redesigned and relaunched as a gaming site.)

By 2005, things had really heated up. Myspace, which had been founded in 2003, began picking up steam; at one point in the early 2000s, it had surpassed Google in terms of numbers of unique visitors. The company generated a healthy $800 million in revenue in fiscal year 2008, but an 800-pound gorilla was about to move into the social networking space.

The Birth of Facebook

Facebook was founded at Harvard University in 2004 by Mark Zuckerberg and some college friends and roommates. Originally limited to Harvard students, the network soon expanded to other schools in the area, and eventually, to other colleges and high schools. (Pretty soon, it seemed that *everyone* was on Facebook. These days, there are probably houseplants with Facebook accounts.)

Remember how the early social networks were counted as successful when they claimed memberships in the millions? As of 2014, Facebook had about 1.28 *billion* users around the world, and those users "shared" 4.75 billion items every day. The average Facebook user spends about six hours or so on the network every month. The site collects 3.2 billion "likes" and comments *every day*.

To put it plainly, the world has never seen anything like Facebook. For better or worse, Facebook is a cultural and economic force to be reckoned with. We're talking about a company that gives away a product for *free*, yet manages to make about $15,000 per minute—and almost $3,000 of that is pure profit. How can that happen?

Well, *you* are the reason it can happen. And me, and everyone else who willingly signs on to Facebook or Google or other such information purveyors.

As has been said before, you're not one of Facebook CEO Mark Zuckerberg's "customers." You are, in fact, the *product* he's selling.

By definition, businesses are meant to generate a revenue stream. The way that Facebook (and Google and Twitter and Instagram and the like) generate revenue is by selling ads—surely a conventional and acceptable time-honored way of making money. Newspapers have always done it, radio does it, and television certainly does it. And we've always been fine with it; if the ads kind of mess up the page layout, well, we've learned to deal with it, and it's kept our newspapers affordable and our television free. (And as one of my crusty old editors once grunted, "Remember, the only reason your copy is there is to keep the ads from bumping into each other!" Note that, young or old, all editors are crusty.)

And of course, the other big social network players—Pinterest, Instagram, Foursquare, Twitter, etc.—several of which are threatening Facebook for dominance as fickle younger users look for something better, more immediate, and of course much cooler than Facebook—survive the same way: They sell ads, and companies will pay dearly to get their products in front of this many eyes. (Many, many eyes. Twitter claims 271 million monthly users, and about 570 million or more tweets are sent every day. That's about 8,000 tweets every second, up from about 5,000 tweets *per day* in 2007. In the meantime, Facebook is shedding teenaged users as they decamp and head for alternative social networking sites that are perceived as being aimed more at younger users. While 72 percent of teens surveyed in the spring of 2014 said they used Facebook, only 45 percent felt the same way that fall. Then again, the teens' parents have only just discovered Facebook and are flocking to it in droves, so Mark Zuckerberg is in no immediate danger of being evicted from his home.)

But there's a big difference between the way traditional media sell and place ads and the way that social networking sites sell and place ads. The monetization of a social network is accomplished by collecting data about *you*. A surprising (and perhaps frightening) amount of data.

Security expert Bruce Schneier has commented that surveillance is in fact the business model of the Internet: "We build systems that spy on people in exchange for services. Corporations call it marketing," he told attendees at a Boston security conference in 2014. "Data is currency, and consumers are willing to hand over their information in exchange for 'free [services] or convenience.'"

Security expert, author, and blogger Bruce Schneier notes that we give up privacy for the sake of convenient communications. *Photo by Ann De Wulf. Image courtesy of Bruce Schneier.*

That data is out there, whether you know it or not, and regardless of whether you approved its release. Keep in mind that social networks don't exist for *us*. Generally speaking, there's no egalitarian, altruistic mechanisms involved in the creation of such tools; they exist to make money, and they do it by collecting information about you and me and selling that information to people who can use it to create targeted, productive advertisements. It's actually very simple: Social networks sell us to marketers who then attempt to sell products to us.

The Trade-off

Is that a fair deal? Millions of us must think so, because by signing up with Facebook or Pinterest or Twitter, we agree to their Terms of Service, or TOS. (We rarely read those terms, of course, which may be what they're counting on. After all, Facebook's TOS is some 4,500 words long, and contains links to at least ten *additional* policy documents.) Buried in the TOS, you can be sure, is language that gives the social network (whether it's Facebook, Twitter, or some other such tool) the right to use our names and images, to collect data about us, and to share that data with advertisers.

What do we get? We get the convenience of having a set of tools that allow us to post our thoughts, read the posts of others, and to share information with friends, acquaintances, and, sometimes, with complete strangers. For that, you get a way to be . . . well, *social* with your family, your schoolmates, and others. In general, this is not a bad thing; many of us enjoy keeping in touch with friends and family in this fashion, and occasionally the ability to connect with others is quite helpful, sometimes literally a lifesaver. (We'll talk more about some positive uses of social networks momentarily.) And for that convenience, we're willing to sacrifice a certain amount of privacy.

The Hidden Parts of the Deal

The reason that we are, by and large, satisfied with this trade-off may have to do with our belief that it's okay to sacrifice a bit of privacy for the sake of convenience. But that belief could be founded on two things that in fact may turn out to be untrue.

First, we believe that the invasion is minor and that the consequences of that invasion are either beneficial or of no import. Neither of these is necessarily the case, because unseen actors are utilizing the data gathered for unknown purposes. Therefore, it's not really possible to determine the extent of the invasion or whether it will ultimately prove harmful, or in what way; you can't analyze (let alone approve) an act of which you know nothing. And besides, in spite of users' protestations that they have "nothing to hide" and don't much care that the world knows about the wild parties they attend or the offhand racist or profane remarks they may have made online, the fact is that one cannot know the future; they may very well have reason to regret their youthful actions in a future they cannot divine.

Second, we believe that we've *knowingly* assented to this brokerage in the first place. But in fact, our assent is often either not fully informed or was not requested at all. Much of the time we don't know that we've agreed to an invasion of privacy, nor have we any idea of the extent of that invasion. Have you actually read Facebook's TOS? Of course not. Do you know how many people Facebook (or Twitter or whatever) shares your information with? Or who *those* people share it with? Do you know exactly what the information consists of? Do you know, for instance, that it's possible to take the "meta-data" collected by such services and use it to construct a map of the locations of your recent physical activities and travels?

On the downside . . .

How to Get Hacked

Chris Hadnagy is a security expert. He's a pen-tester (a penetration tester), a white-hat hacker who breaks into companies' networks at the behest of company officers who wish to discover weak points in their corporate security, but he generally doesn't get in programmatically. He's a social engineer; he uses a disarming smile, his easygoing manner, and his knowledge of people's foibles to get you to give up the keys to your (or your boss's) kingdom. And he uses Facebook. Facebook and Twitter and the like—the Internet in general—are his best friends, and he can use them to find out what he needs to know about you in order to get you to help him compromise your network.

> As he says, "I make a living by teaching companies how vulnerable their human network is."
>
> Social networks have made his job much easier. With them, he can find exactly the piece of information he needs in order to set up some pretext that will have you happily handing over the log-in or other info he needs to compromise your work network.
>
> "Take Twitter," says Hadnagy. "Only 140 characters, but people tweet *everything*! They tweet their geolocation, what they're eating for lunch, where they work out, pictures of themselves on vacation—they tweet everything. LinkedIn has your whole corporate history: your résumé, your skillset, your clearance levels, what you've done, what you are doing, what experience you have. And Facebook is everything *else* about your life. If you don't have these social networks locked down and they're public, then you are an open book to a social engineer."

How to Undo a Social Media Mistake

People of all ages use social media, and many of them post things that they'll eventually regret. Sometime after you post to Facebook that negative comment about your boss or your brother-in-law, or sometime (perhaps years) after you post that image of yourself with the red Solo cup in your hand, you may have occasion to wish you hadn't done so.

Unfortunately, the Internet is forever. Even if you "remove" the poorly considered post or the incriminating image, it's still up there, cached or shared or copied. You can't go back.

So, are you doomed to live with a mistake you made simply because you had too many margaritas or too few life experiences?

Well, not necessarily, says security expert Chris Hadnagy. Hadnagy, the author of *Social Engineering: The Art of Human Hacking* and *Unmasking the Social Engineer: The Human Element of Security*, points out that you can, in effect, social-engineer your own information. The answer, he says, is to *make more good news than there is bad.*

"In other words," he says, "if you've got one bad picture out there that was taken in September of 2014, and for the next two years every week you have a *good* picture, a good story, something great about your career, your life,

Author and security expert Chris Hadnagy earns a living using social engineering skills to break into corporate websites. Because who wouldn't want to do that? *Image courtesy of Chris Hadnagy.*

your education, they're gonna have to dig really far to find that bad picture. But if you get so nervous that you stop using Facebook or whatever, and that's the last thing on there, then that's probably the last picture that people are gonna have of you floating around on the Internet."

The Three-Edged Sword

Social networking is something of a three-edged sword. The first edge is presumably the reason we use it in the first place: There truly is a genuinely "social" component to it. It helps us keep in touch and learn about our friends and colleagues, and it provides an opportunity to use our friends' networks to make new friends of our own. It's a communicative tool that can connect—and then connect those connections, creating a web of, if not friends, then acquaintances, at least, that grows geometrically. You can have *a lot* of friends on Facebook or Twitter or what-have-you.

The second edge is commercial: Many of the participants in social networking have no interest in "social" contacts at all; they're there to make money. (Of course, the purpose of the entire platform is to make money, so that's hardly surprising.) They post material not to inform or to entertain, but to sell something: a product, a service, themselves. Selling something is the whole reason they participate at all.

And that's not necessarily a bad thing. Often they're very up-front about it: "Hey, we're Axel's Flowers. We're pretty cool. Look at this pretty picture of the pretty girl holding the pretty rose. Nice, huh? Well, you can be just as cool. Come buy flowers from us." Nothing wrong with that, really. As we've noted, their ad dollars are what support the (free) social networking platform you're using; without them, you'd have to pay for the service. (Of course, some merchants are a bit less up-front about it, but you usually know that you're looking at an ad or some sort of "sponsored post.")

The third edge, though, is a dark one. It comprises those who are either outright scammers or the next thing to it. They post "clickbait" (sometimes in the form of glurge-filled, sappy stories or videos) or a photo of an old-fashioned metal lunchbox adorned with a photo of an early-era TV character and ask you to "Click here to 'Like' if you remember this!" Well, of *course* you remember it; anyone who was between five and fifteen in 1964 remembers Lassie and The Beav and Ed "Kookie" Burns. All 1990s kids remember the Power Rangers and Buffy and the young men and women of *Dawson's Creek*. Most of the time, the company is just garnering meaningless "likes" and clicks so that they can charge their advertisers more, or, worse yet, so they can turn around and sell the page.

At the very bottom of the "clickbait" barrel are those who purposely mislead viewers while at the same time profiting from someone else's misery. You've seen those posts. They show a photo of a sickly or wounded child and ask you something along the lines of "Amanda has a terminal disease and has only weeks to live. Let's brighten her day. How many 'Likes' can we get for poor Amanda here?" Well, who wouldn't want to brighten this child's (or those parents' or that soldier's) day? And so easily, too; it only takes a mouse-click and costs you nothing!

But it's almost always a clickbait scam. The photo is an old one, normally used without permission, and it may not be of Amanda at all. Someone simply wants our help in increasing the value of their page for whatever reason. (Which is truly a shame, because there *are* "Amandas"—and also humanitarian organizations—out there that do deserve our attention, our love, and even our money. Sadly, those good apples often get lost among the hundreds of bad ones in the social networking barrel.)

Founder Brett Christensen's Hoax-Slayer website garners between 1.5 and 2 million page views per month, while his newsletter goes out to around six thousand people. *Image courtesy of Brett Christensen.*

Brett Christensen and his two sons run Hoax-Slayer, a website devoted to debunking these sorts of bottom-of-the-barrel hoaxes and scams. Christensen, based in Australia, deals with these scams and hoaxes day in and day out; some are just silly, but some make him truly livid.

"Some of them make me angry, especially ones that are racist, designed to incite bigotry and hatred, or promote a political agenda," he says. "But I think the ones that make me most angry are the 'sick baby' Facebook hoaxes. These people steal pictures of sick or injured babies and claim that Facebook or another company will donate money to help the child each time the message is 'shared' or 'liked.' They do this just to get page 'likes,' because Facebook pages with high 'like' numbers can later be sold on the black market or used to distribute scams or spam. These sorts of hoaxes can cause great distress to the families of the children pictured; in some cases, the child has already died."

It's unfortunate that some people have to subvert a truly impressive and potentially useful technology, using it either to attack people or line their pockets at the expense of others.

Tools of the Trade

Technology, as helpful as it might be, can be used against you; it's as simple as that. People, for whatever reasons, have access to information about you that you did not realize was out there for the taking. On the plus side, you too have tools that you can use to fight back and to monitor your own social media footprint. Knowledge, after all, truly *is* power.

Take Immersion, for instance. Immersion is a tool built by the MIT Media Lab. It examines your e-mail patterns and builds visual representations of those patterns that can be used to see what information could be gleaned by someone who had access not to your actual e-mail conversations, but "only" to the metadata—the information *about* the e-mails: who sent them, when, to whom, how often, and so on. Metadata sounds harmless (and indeed, the NSA has made a practice of responding to accusations of domestic spying by countering with, "Well, we're only looking at metadata, not the actual e-mails or conversations"), but metadata can tell you a lot. In fact, as security blogger Bruce Schneier has pointed out, "Metadata is far easier to store, search, and analyze than actual content, and actually has far more value to an intelligence agency. Metadata is fundamentally surveillance data."

Immersion uses e-mail metadata to build visual bubble maps that show whom you've been e-mailing, when, and how often. Because it's an educational tool and not actually malware, Immersion will ask your permission before examining your data. You simply log in with your Gmail, Yahoo, or other e-mail account, and let the tool examine your e-mails. Remember, it's not looking at the content of those e-mails, just the metadata: header info, TO: and FROM: addresses, etc.

Data of this sort can reveal far more than one might think, even though it's "only" data about data. The graphs that Immersion creates can reveal clusters of interrelated groups: who speaks to whom (and when and how often, as well as who recently *stopped* speaking to whom), and it can uncover the relationships between those social groupings. *Boston Globe* writer Abraham Riesman called it "a cloud of knowledge about your behavior that, once you confront it, can literally change your life."

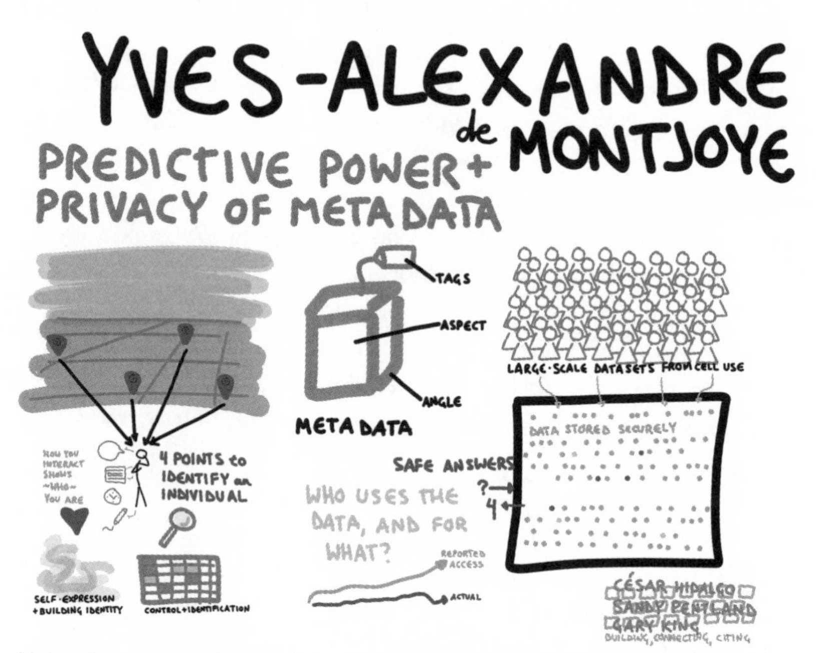

Metadata in action. Information *about* information can, especially in large datasets, be used to predict actions and identify individuals. *Image courtesy of Wikimedia Commons user Willowbl00; image licensed under the Creative Commons Attribution-Share Alike 4.0 International license.*

There are other such tools, some more intrusive than others, but almost all of which rely on information that is freely available. These tools are not finding "secret" information; they simply uncover, aggregate, and serve up information that's already there.

Echosec is one such tool, and it's the application I was using when I was looking to see who was posting what at various locations, including Pyongyang, North Korea; the White House; and various bars in my hometown of Lincoln, Nebraska.

Using Echosec to uncover the contents of social media posts is simple: The application displays a Google map; just set a date range and indicate a geographical location on the map. The social networking feeds immediately start filtering in below the map. Even using the free demo version, you can see posts from Foursquare, Twitter, Flickr, and Instagram; behind the paywall are more feeds, a more-sophisticated interface, and some additional functionality.

Again, none of this information is secret. None of it is private. It's out there—it's just not been terribly accessible. Until now.

"We're just a different type of search engine," says Echosec CEO Karl Swannie. "The information is already there, and you can find it through the individual service providers. We're very much like a Google News, where we just kind of aggregate information into one single view."

Swannie is very open about what his company is doing. "We work with police, we work with other organizations to let people know, to educate people, so the digital literacy side of what we do is very, very important, telling people that this is out there."

Echosec obviously has law enforcement applications; the police would very much like to know, for instance, who was tweeting what in the area immediately surrounding a crime scene at the time a crime occurred. Echosec lets them see that—even days or weeks later.

But the tool is also used in other areas, says Swannie.

"We're being used in high schools for anti-bullying," he says, "and just for watching over those areas. Usually it's by police departments and people that are appropriately trained. And they're also [using Echosec to] go into those high schools and teach the kids what is appropriate and what is not appropriate to post."

One of Swannie's main points is that security matters: "If you're in a fenced building or in an area that has some sort of security, don't post anything! If you're on a military base, for God's sake, stop posting stuff! It seems like everybody who's on a military base, the first place they check into is the ammo dump. And I can show you almost every military base across the United States, and even globally, where people are posting from those locations."

The message here, notes Swannie, is that if an entity cares enough to erect a physical fence around its environment, it also needs to have given some thought to erecting a *digital* fence. Echosec helps to point out that need, so that employers—private or government—can take steps to ensure that private information stays private.

Jason Jubinville is another Echosec spokesperson. He points out that there's also a humanizing aspect to being able to peer through the social media lens in order to really see what's happening in distant places.

"If you go to a conflict area," says Jubinville, "you can go there via a news story on CNN. But if you go there via social media—if you go to South America, or to Nigeria, where the bombings were—all of a sudden you see the panic, you see the stress, you see all of the atrocities, and it really just kind of hammers home that they are human and that they're *not* just a story, they're *not* just text on a web page."

Echosec is being used to aid public safety, improve corporate security, and combat insurance fraud. Swannie's message is simple: This data is already out there, so don't shoot the messenger. And while it can be gathered for crime prevention and other positive applications, it can, of course, also be misused.

Karl Swannie, CEO of Echosec, notes that the company's technology is being used by schools, corporations, and law enforcement. *Image courtesy of Karl Swannie.*

Note that it is possible to avoid being swept up in the Echosec (or similar tools') net: Just turn off geotagging on your social media accounts. Since Echosec relies on your location, it cannot track your posts if your location is unknown. (Then again, geotagging is the entire *purpose* of some social networking apps, including Foursquare, Shopkick, or Foodspotting.)

The Positive Side of Social Media

Of course, it's not all doom and gloom. While your (supposedly private) conversations with your friends can hurt you later, and although data about you can be mined for marketing purposes, social networking has also been—and is currently being—used for good.

Pro-democracy demonstrators in Hong Kong used Twitter, Instagram, and other such tools to coordinate their activities, to get their message out, and to show the world—often immediately and viscerally—how they're being treated while attempting to demonstrate peacefully.

During the Iranian election protests in 2009, Twitter, Facebook, YouTube, and blogs were used to upload information for worldwide consumption.

In the 2010 "Arab Spring" revolts that resulted in the turnover of power in several countries, and in sustained demonstrations in many others, social networking proved to be a powerful tool for collective activism.

Such applications are engines for engagement, though whether they're driving protest or merely documenting it is not always clear. And it's important to note that the authorities can also use social media and other such digital tools, and can sometimes monitor their use by protestors. This includes even supposedly "off the grid" social networks such as that utilized by FireChat, which proved popular during the Hong Kong demonstrations. FireChat uses a dynamically created "mesh network" that is essentially peer-to-peer: Two people can connect directly, without using the (easily monitored) Internet. If more users join the network, its reach expands with each new addition. Eventually, hundreds or thousands can be chatting on an ephemeral ad hoc network that will disappear as people leave it. From the protestors' point of view, the problem with FireChat is that *anyone* can join the network, including the authorities. (When we say that technology tends to democratize, keep in mind that it does so for *everyone*, regardless of which side of the political fence they're on.)

Social Media for Social Good

Karl Swannie, quoted at length earlier in this chapter, points out that his social networking search tool, Echosec, has been used to help combat poaching in Africa. For instance, poachers these days often post information about their successes, even going so far as to tweet photos of dead elephants and rhinos. When tourists come across the carcasses (generally intact except for the missing tusks, horns, or other valuable parts), they too can post, alerting authorities of their finds. With a location now locked down, those game wardens and police forces can use Echosec to backtrack and find social networking posts related to the poaching activity. This often allows them to find the perpetrators.

In May of 2014, floods in Serbia devastated the Balkans. Several dozen people died, and entire towns were inundated. Railways that could have been used to deliver aid to the victims were themselves damaged or submerged, and some bridges were wrecked.

It was a mess, and people were suffering; the Serbian government, naturally, appealed to citizens for help.

In response, Twitter users launched a website to coordinate relief and distribution efforts. Support poured in, and social media users kept victims and aid workers alike informed.

Something similar occurred in the Philippines when Typhoon Haiyan (also known as Typhoon Yolanda) hit in November of 2013. Social media helped show the world what was happening. (What was happening was devastation on a huge scale. With Haiyan packing sustained 175-mile-per-hour winds, it simply destroyed cities and towns wherever it touched down, leading to widespread looting and lawlessness. In the end, some five thousand people died.)

Again, social media helped show the world what was happening in the disaster zone, leading to an outpouring of help of various kinds, financial and material.

In 2003, webcomic writers Mike Krahulik and Jerry Holkins created Child's Play, an organization devoted to providing hospitalized children with toys and games. Using social media as a messaging tool (the two cofounders have massive followings on Facebook and Twitter), the organization has funded many millions of dollars' worth of toys since its inception. (In 2013 alone, the group raised over $7.5 million.)

An especially evocative social media undertaking is the It Gets Better Project. Aimed at LGBT teens and their supporters, the project's goal is to let LGBT teens know that they are loved, that people support them, and that bullying will not be tolerated. (This is particularly meaningful when one considers that social media itself is often used to bully. There have been instances in which young LGBT social network users—and others—have been driven to suicide by those who taunt them over the Internet.) The organization's Facebook page has been "liked" well over 350,000 times, and its Twitter account has attracted 114,000 followers.

Formed in 2010 by domestic partners Don Savage and Terry Miller, It Gets Better is spreading a powerful message of support for those who most need it, and it's using social media to do so. The project encourages participation by gay and straight celebrities of all stripes, and the project's website notes that it has received support in the form of videos from such luminaries as President Barack Obama, former Secretary of State Hillary Clinton, Representative Nancy Pelosi, Adam Lambert, Anne Hathaway, Colin Farrell, Matthew Morrison, Joe Jonas, Joel Madden, Kesha, Sarah Silverman, Tim Gunn, Ellen DeGeneres, Suze Orman, and the staffs of The Gap, Google, Facebook, and Pixar.

Or consider the plight of women, many of whom live in more-or-less constant fear when out alone, even in their own neighborhoods. As one *Washington Post* writer noted, women give up a lot simply because of fears they shouldn't have to deal with in the first place: "We forgo a nighttime event because we don't want to travel home alone afterward. We forgo an evening jog because running at night is a luxury only men possess. We forgo a comment or an outfit or a friendship because it might imply an invitation we don't wish to convey."

According to the National Opinion Research Center's General Social Survey, women are more than twice as likely as men to say that they're afraid to walk in their neighborhoods alone at night.

Obviously, social media is not capable of solving the problem, but it can serve as a tool to help start a discussion about the issue. And as part of that, the trending #YesAllWomen hashtag serves to link together participants in that discussion (pro *and* con), allowing women to share their stories about harassment, and women and men to respond to those stories. In effect, it's become a national (or even worldwide) sounding board. And even if the discussion occasionally generates a certain amount of rancor (which it seems to, based on comments on the site), isn't that better than the issue not being discussed at all?

The Arrival of the Technology-Based Social Network

We've always had "social networks," of course—groups of friends clustered into what amounts to discrete networks; there were (and are) work friends, school friends, family, friends associated with leisure activities, and the like. And though there was often some overlap, those groups stayed largely separate. (Which was often a good thing; after all, as much as you love her, do you really want your blue-haired great-aunt Ruth to be privy to every discussion you have with your school chums? Didn't think so.)

But now technology—specifically, the Internet—underpins our social networks, and it brings to them a breadth and speed and reach heretofore impossible. Facilitated by the incredible communicative power of today's sophisticated hardware and software, our tech-based social networks allow us to create huge consortia of associates of various types, to do so very quickly, and to reach out to friends-of-friends-of-friends much more easily than ever before, thus greatly expanding our social reach. Our social connections are expanding at such a rate and to such a degree that one wonders just how tenuous some of these links really are, and if they're actually worth having. (Of course, among young people, the sheer number of friends or followers is something one endeavors to increase, because a larger number of friends is perceived as making one appear more popular—and if you remember your teen years, being popular was incredibly important to most of us.)

Social networking in this technology-enhanced form has not been around for a terribly long time—perhaps about twenty years, according to the brief recap earlier in this chapter. It could not exist without the Internet, or something very much like it, and, as noted in an earlier chapter, the Internet was one of those technologies born of (real or perceived) military need. As is the case with much of the technology we're discussing, only much later was this level of connectivity available to civilians.

Even then, the cost of early computers and accessories tended to restrict availability. In 1982, according to the Brookings Institution, acquiring the computing power of an iPad 2 would have cost more than 360 years' worth of wages. A bit closer to home, a 1993 Dell XPS P60 system would have cost over $4,000 in 1993 dollars; today, a Dell Inspiron with significantly more memory, storage, and computing power is less than $500 at your neighborhood big-box electronics store.

In 1986, two partners and I started an educational software company called StudyWare. The company created what was for its time some fairly sophisticated test-preparation and course-review software. At the time, we would tell people, "If you provide your child with a computer, you will be giving her an advantage." Only a couple of years later, we had changed our tune: By the early 1990s (by which time the company had been acquired by Cliffs Notes), we were saying, "Look, if your child does not have a computer, then she is at a decided *dis*advantage." The standard had shifted. Now it was simply *assumed* that students would have computers; if not, they were handicapped. It wasn't just marketing, it was (and is) actually true, and the reason it was true was that the cost of computers was plummeting, even as the machines were becoming more and more powerful. (A shift that continues today, of course.)

Soon, the prices had come down to the point where it was no longer only the children of wealthy parents who had computers, it was also children of the merely well-to-do. And then of the solidly middle class. Now, while we cannot say that all children have access, the digital divide in the United States continues to narrow: As of 2008, some 85 percent of households with eight- to eighteen-year-olds had personal computers, and those numbers have risen in the interim. The Child Trends DataBank notes that, in 2001, about six out of ten children access the Internet from home. (Sadly, web access does still correlate to household income, though not nearly as much now as it used to; well over half of the children in households earning less than $15,000 per year have access to a computer at home.)

The standard is different in third-world countries, but the same sort of curve can be seen; Africa is the fastest-growing smartphone market in the world, and as Voice of America has noted, "The rise of smartphones has given millions of Africans Internet access for the first time."

And with almost ubiquitous access to the Internet (especially when we include access provided by mobile devices of various sorts) comes the ability (for better or worse; the jury is still out) to participate in a wide variety of social networks.

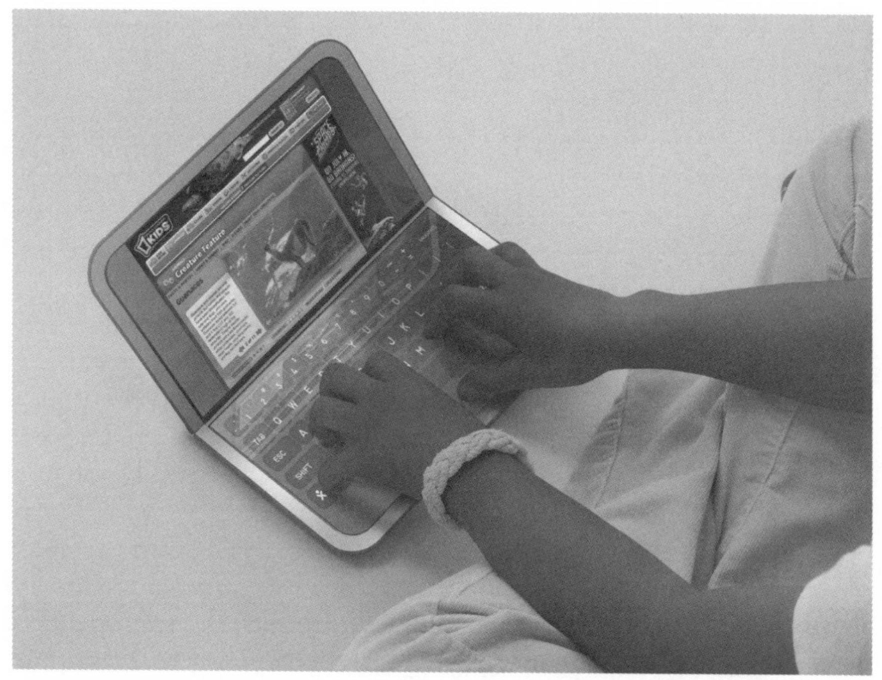

The rates of home computer and Internet access continue to climb, though serious gaps remain. This photo illustrates part of a design study undertaken to find ways to narrow those gaps by providing computer access to third-world residents. *Image licensed under the Creative Commons Attribution 2.5 Generic license. See http://wiki.laptop.org/go/OLPC:License.*

Will Your Refrigerator Have a Social Network?

It's not just teens that are talking to one another and sharing info over the Internet. More and more, previously unconnected devices ranging from your refrigerator to your lightbulbs, and from your car to your home thermostat, are engaged in (what we hope are) friendly conversations of one sort or another. All of your children, friends, and family members are connected, and soon all of your "things" will also be linked. It's called the Internet of Things, of course, and that's where we'll head next.

Chapter Eleven

Everything Is Connected: The Internet of Things

First were mainframes, each shared by lots of people. Now we are in the personal computing era, person and machine staring uneasily at each other across the desktop. Next comes ubiquitous computing, or the age of calm technology, when technology recedes into the background of our lives.

—Mark David Weiser

The Well-Connected Pig

Whenever and wherever the so-called Internet of Things began, it was almost certainly not in 2006 on a pig farm in Holland. But that's where and when it may have gotten just the tiniest bit of a nudge. That was the year in which a Dutch pig farmer determined that he really needed to do something about a vexing problem that had bedeviled him for many months: The pigs, which were watered by a system consisting of nipples connected to pipes, kept damaging the nipples, chewing them off and allowing water to run freely into the pens in which the pigs were kept. Worse yet, this seemed to happen most often on Friday evenings, just after everyone had locked up and gone home; thus, thousands of gallons of water poured out into the enclosures over the weekend, draining into a reservoir beneath the pens.

Because of the construction of the pens and reservoir, it wasn't terribly messy and the pigs were certainly in no danger, but it *was* expensive. The farmer was buying water cheaply enough, at about fifteen cents per ton, but pumping out the slurry that resulted from the flooding was costing more like twenty dollars per ton; since the water was pouring out at about 1,000 gallons per hour, this quickly added up to quite a tidy sum.

The farmer called Dosing Solutions, a UK company that specialized in water-flow monitoring, to ask them if there was a way to monitor the pigs' water such that he would know if more water than normal were suddenly being used.

There was. In fact, monitoring and collecting information about water and similar issues became such an important aspect of its business that Dosing Solutions spun off a new company, General Alert. The newer company concentrates on collecting and transmitting data about various agricultural systems, largely in the poultry and pig markets. Suddenly, there was an awful lot of data being collected about pigs and chickens. Eventually, actual temperature transponders were inserted subcutaneously into the pigs; the pigs themselves were now transmitting data about their own health and status. The Internet of Pigs had been born.

"If the Internet of people is everybody having an e-mail address and sharing their innermost thoughts on Facebook," says General Alert managing director David Welch, "the pigs sort of do the same, really."

As with humans (and old automobiles), pigs and chickens usually exhibit various warning signs before they are actually, noticeably sick. One of those signs is an unusual reduction in water consumption.

"They don't feel right," says Welch. "They can't be bothered to go over to the drinking areas, and because clever algorithms are at work compiling and graphing the animals' daily water consumption, if the pigs or chickens should be growing and drinking more every day and the data suddenly shows that today's consumption is *lower* than yesterday's, the system can alert you and you can act on the information, saying, okay, this group of pigs is sick. Why don't we treat them or separate them or give them extra vitamins or something while they're sick, in a very small way? Doing so could actually save quite a bit of money and heartache later on by catching the problem early."

These days, General Alert monitors more than just water flow, of course. The company has sensors in the field that track temperature, humidity, fog density, carbon dioxide, ammonia, and other such chemical and environmental indicators.

David Welch is the managing director of General Alert, a UK company that specializes in collecting and monitoring water flow and other environmental data. *Image courtesy of David Welch / General Alert.*

Welch notes, "The driving force behind this is society's desire to reduce the antibiotic input into animals. We sort of want to keep antibiotics for our own use, and we don't want to see resistance built up. Now, in a perfect world, you would never give antibiotics to animals, but you sort of have to look after them, because nobody wants to eat diseased animals, and nobody wants to see animals suffering. So treating in a timely manner using information derived from the sort of data we're collecting puts you in a very strong position to actually do something."

And we're talking about a lot of data. Sensors deployed by General Alert, a relatively small company in England concentrating almost exclusively on the agriculture industry, are now generating more than 300,000 data points a day, and Welch says that the number is doubling every month or so.

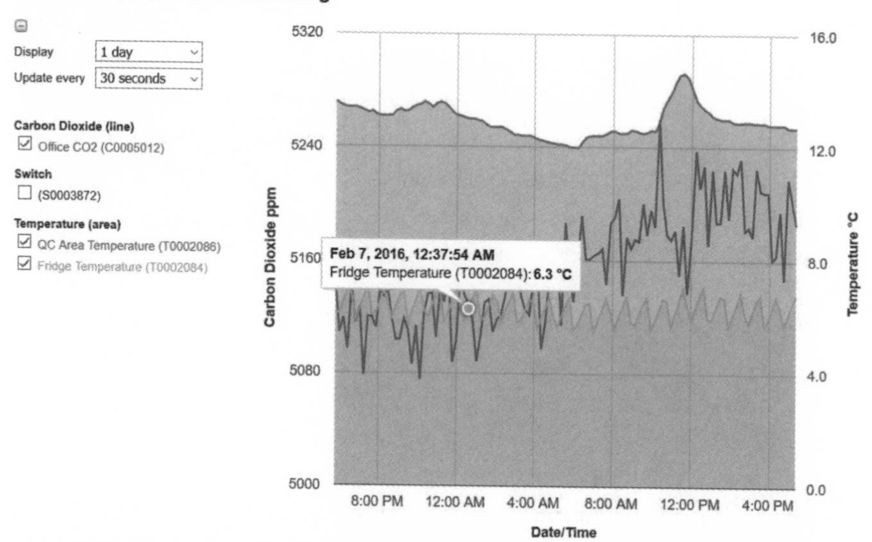

UK-based General Alert's sensors can track a number of environmental variables. This is the readout from sensors that measure conditions in David Welch's office. It's obvious that David was in his office early in the morning, and that multiple visitors were present around 11:30 a.m., since both the temperature and carbon dioxide levels jumped. Importantly, the temperature in David's refrigerator stayed pretty constant. *Image courtesy of David Welch/General Alert.*

In fact, there's enough data coming in that General Alert farms out the crunching of that data to 1248, a Cambridge-based company that specializes in Internet of Things (IoT) strategy, database design, and application development. There's so much data that making sense of it requires that someone specialize in this sort of analysis.

"That is a function all of its own," says Welch. "You can't do that sort of stuff on your own PC in the office. So we've had to rope in these guys to say, 'We're struggling with this amount of data. Can you keep scaling it up as we need it?' "

Big data. Really big. And really fast.

Welcome to the edge—just the very edge—of the world of big data. Forget about all of the terabytes and petabytes of info created daily with which we're currently dealing; data from the IoT will dwarf that seemingly colossal stream of information. At the rate the Internet of Things is growing, analysts anticipate that by 2020, anywhere between 26 billion and 50 billion devices will be connected, and those numbers don't necessarily include increasingly popular wearables such as smart watches and intelligent tennis shoes and moderately clever pocket lint. The IoT will collect and transmit only small snippets of data at a time, but it will do so billions of times per day, all day and all night. General Alert's current 300,000 data points per day won't even be a blip in that data stream.

The IoT will generate incredible amounts of data at very high speeds, and the expectation is that we will be able to make use of that data in real time. The fact that my fridge's temperature fluctuates by 7 degrees over a calendar year is interesting, and perhaps even important; but if the temperature rises by more than a few degrees on any given day, I really need to know that *right now*. Similarly, a farmer needs to know the current moisture content in a field and whether that content is likely to be increased by an upcoming rain or reduced by an anticipated drought, and he needs to know that now, not in a day or a week or a month, after the data has been collected, tab-

According to Nest, by September 2013 (after only a couple of years on the market), "smart" Nest thermostats had saved a billion kWh—enough energy to charge all the smartphones in the United States for two years. *Image courtesy of Nest, Inc.*

ulated, and analyzed. Likewise, a connected fire detection system needs to sound the alarm and contact the nearest fire station immediately, not sometime after the data has been cleaned, parsed, categorized, and transferred through multiple servers around the world.

It's an incredible amount of information being generated, mostly in tiny data points that must be collected and sifted through, but much of it must be actionable immediately. The Internet of Things is nothing if not immediate.

The Big Issues: Compatibility and Privacy

Consider what the Internet of Things really means: At its core, it's simply, as one writer deftly summed up, "a wide-ranging ecosystem of everyday physical objects connected to the Internet, capable of identifying themselves and communicating data to other objects on the network."

But the fact that the concepts behind the Internet of Things are simple to grasp should not delude us. In its implementation, the IoT is neither simple nor secure.

Part of the problem is that, unlike when Vint Cerf, Bob Kahn, Steve Crocker, and others were designing the Internet itself, including protocols that would standardize its use, there is no "cop on the beat" here, and thus, no standardization. Service providers and manufacturers such as Sony and Intel are free to develop their own protocols, specifications, and "standards," and hope to force others to develop with those standards in mind. (Because that worked so well for Sony when it came to those Betamax videotapes. And the Sony MemoryStick. And DAT. And . . . Well, anyway, surely Sony has learned its lesson by now, right?)

There are currently at least six competing "standards," some proprietary, some open-source. Apple, Philips, Texas Instruments, and other large companies are behind the HomeKit standard. Technology giants such as Intel, IBM, and Cisco are rallying behind the Industrial Internet Consortium. And Sony, Samsung, Dell, McAfee, and others are behind the standards (many of which are yet to be defined) created by the Open Interconnect Consortium.

The competing standards may cause no end of trouble for consumers, but that lack of standardization has in fact created something of a cottage industry for Cloud-based companies that promise to "broker" those connections, freeing manufacturers from the drudgery of attempting to find the "one best way" for the sensors on their devices to connect to the Internet and to other devices.

One of these is Redwood City, California–based Arrayent. The company works with businesses such as Whirlpool and FirstAlert that produce devices they would like to connect to the IoT, but which may not have (and may not wish to develop) the programming and technology expertise needed to build that level of intelligence into their devices. Arrayent solves the problem by providing a Cloud-based system that brokers those connections, embedding a bit of software in the products and providing a system that helps out the manufacturers.

"Up in the Cloud," says Arrayent founder Shane Dyer, "there's essentially an agent for each of these devices that's used to talk to things like mobile applications and to their service departments and other cases like that. We provide that platform to help these manufacturers, and these retailers make the jump from unconnected products to connected products."

Providing that service, says Dyer, streamlines production and allows manufacturers to simplify the development effort and keep the unit costs down.

Dyer notes, "If I'm buying a product, what's not happening is people waking up on Saturday morning and saying, 'Oh, I have a home automation product problem.' Instead, there's probably only one or two things that are top-of-mind for you, and you're going to go to the hardware store or you're going to go to Walmart and buy one product, and maybe you'll spend a little bit more money to buy the one that comes with an app. I think that's the way the mass market's going to encounter the Internet of Things. So it's really important that not only must these things be really simple, but they also need to be fairly low-cost."

In general, the bottom line (in more ways than one) is that if the IoT is to make any headway, and if people are truly going to look at ways to automate their homes and connect their devices, the manufacturers must find ways to make doing so simple and affordable. Otherwise this whole thing is a nonstarter.

After all, do I really *need* a smart fridge? Of course not. And I could certainly live without a $1,700 smart washing machine. (Honestly, do I really need to get a text message telling me that my laundry is ready for me to move it to the dryer? Now, if the machine were smart enough to move it *for* me, then maybe we could talk. And if it could also fold and iron, well, now I'm sold. But until then? Meh.)

As *MediaWeek* writer David Rowan notes, "At no stage have any of these product marketers answered a simple question: Just because a device can be packed with sensors and put online, does it actually serve a consumer purpose?"

Some of them may serve some purpose, of course; it's just a question of whether that purpose is perceived by consumers as an actual *need* of some kind. And to Shane Dyer's point, if the manufacturers can make the experience simple enough and affordable enough, then the market will go for it. Unfortunately, simplicity, affordability, and functionality are not the only concerns.

Security on the IoT

Whenever you connect a node to a network (in this case, to the Internet, which is in fact a network of networks), the possibility exists that you have also exposed a potential weakness—that your threat landscape has been broadened and your vulnerability thereby increased. When your entire home, your car, and every device you own down to your watch, washing machine, and wallet, are connected, any or all of them could act as portals for malicious actors of whatever sort, whether that's black-hat hackers, cyberthieves, or government agencies. The risks to which you expose yourself merely by connecting your computer to the Internet are multiplied a thousandfold once you start connecting your household and mobile devices—not to mention your car and your home itself—to one another and to the Internet.

This may be another area where we either have to come up with secure IoT protocols, build the security into the devices themselves, or engage the services of some third party in order to preserve some level of security.

Part of the security problem arises from the fact that in many cases, what we're really doing is putting actual computers in our household devices. This not only raises the cost, but it also provides a bigger target for hackers and eavesdroppers.

"I think a lot of the products that you see out there right now," says Arrayent's Shane Dyer, "are these kinds of big computers that we're putting into our products: a fridge that has a little e-mail server in it, and things like that. We think that that is exactly the opposite approach to what should be done. What

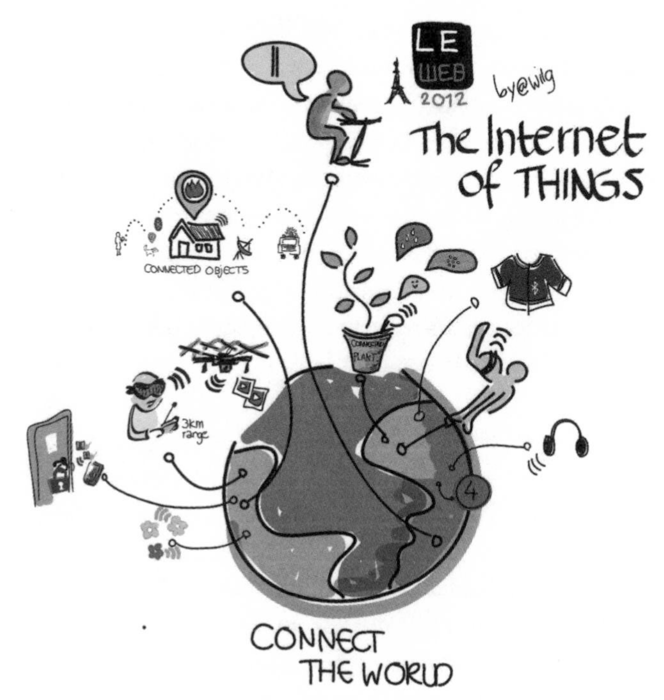

Like it or not, the Internet of Things (IOT) is coming. Everything from your houseplant to your refrigerator will be connected to each other and to the Internet at large. The IOT will be incredibly useful, but critics worry that it may also be incredibly risky, providing new gateways for hackers and cybercriminals. *Image licensed under the Creative Commons Attribution 2.0 Generic license. Drawing by Wikimedia Commons user wilgengebroed.*

we do is try to make the devices themselves very, very simple. You don't want a computer-fridge. You just want a fridge that has connectivity. And then, by making the device incredibly simple with that connectivity piece, and actually putting a lot of the complexity up in the Cloud, it means you've got a fighting chance at going through and securing these things. The targets tend to be much, much smaller than when you have more heavyweight general-purpose computing built into all these devices that are around us."

And the IoT really *needs* to be secure. After all, we're talking about devices that, while perhaps not directly involved in high-security transactions themselves, may be connected to devices or services that *are* involved in such transactions. Think of it this way: In and of itself, your fridge is not much of a security risk. But when it notices that you're low on milk and decides to help out by ordering some from your local grocer, it may connect to a payment system that *is* a security risk. In a connected world, if someone could hack your fridge, they're potentially getting access to much more than just information about your choice of ice cream or a count of how many times your freezer door has been opened this month.

Dyer says that using Cloud-based services that broker such connections is one way to help ensure security, because these services can themselves police those connections.

"Up in the Cloud, where we do banking and things like that, there could be teams of security professionals working on making sure that that's secure," he says. On the other hand, "There's probably not a security professional that's gonna run around your living room and make sure your smart sockets and your thermostat and your other things are kept secure and up-to-date."

According to some technologists, part of our IoT security problem (as well as several other looming problems we've not yet discussed) has to do with the fact that full-blown Internet protocols are being utilized to enable communication with the Internet and with each other. Some think that this approach is overkill, completely unmanageable, and will lead to serious security issues.

The problem boils down to one of volume and complexity. As technologist Francis daCosta notes, "The quantity of devices in the Internet of Things will dwarf the traditional Internet and thus cannot be networked with current protocols, tools, and techniques."

Simply put, the complex protocols currently in use cannot handle the millions or billions of devices it is estimated will eventually populate the Internet of Things, says daCosta. The sort of overprovisioning we're used to (and which serves us well when enabling complex Internet-based communications) simply will not scale to a vast, extended Internet of Things consisting of simple devices that, generally speaking, only need to send brief blips of data that need not even be terribly accurate; after all, if anything is missed, it will be made up for by the next blip to come along, mere seconds (or even fractions of a second) later.

And when complex Internet protocols *are* used, they carry with them the same sort of security risks we've now been battling on websites and routers for years. In the end, all of the vulnerabilities we've seen in our computers will now also exist in our toasters, refrigerators, and thermostats, unless we find a different approach, say daCosta and others.

There's much more to this, of course. In fact, a thorough examination of potential IoT security issues could fill an entire volume. In the meantime, daCosta's *Rethinking the Internet of Things: A Scalable Approach to Connecting Everything*, though a bit technical, is highly recommended. And be sure to check out the "On the downside . . ." sidebar in this chapter.

On the downside . . .

260 Billion Data Leaks, and Counting

By now we're all pretty used to the idea that connectivity brings with it certain security and privacy risks. Connecting with friends on Facebook can mean that you're unknowingly sharing information with strangers—some out to sell you products, some who may have more nefarious purposes in mind. Sending out Instagram posts or Twitter tweets? You may be sharing your location and other personal data with thieves or bullies. And of course, your Internet-connected computer needs firewalls, anti-malware applications, and other such tools to guard it—and you—from virus infections, identity thieves, and more.

In fact, given that every one of your connections to the outside world is a potential safety or security risk, essentially a window of vulnerability, what do you suppose will happen when *everything* you own, from your shirts to your shoes and from your watch to your wastebasket, is connected? What happens when all of these devices are speaking to one another, with many of them tied via the Internet to other items or to servers that may be in your home, around the block, or on the other side of the world?

What happens is a deluge of data, one which is undeniably useful but also difficult to collect, store, and analyze. Perhaps worst of all, it's a deluge that is also almost impossible to secure.

As attorney/journalist Jeff John Roberts has noted, the Internet of Things has indeed arrived, but the rules needed to run it efficiently and securely have not.

The result? "Unprecedented possibilities for invasion," he says.

Vint Cerf, one of the founding fathers of the Internet, is an enthusiastic proponent of the IoT, pointing out that it can transform everything from business to home life to medicine. But he worries about its security implications: "You could have a situation where Bank of America succumbs to a DDoS [Distributed Denial of Service] attack from one hundred million connected refrigerators in the United States," Cerf said after a security company claimed to have uncovered evidence that "smart refrigerators" had been compromised and used to relay spam e-mail messages.

Of course, it is possible to make IoT devices secure, but it takes time and money—and we're short on the former and unwilling to spend much of the latter. Security professionals such as Jim O'Gorman don't think that providers and vendors will be willing to spend that time and money until a crisis forces them to do so.

"At the end of the day, there's not much that will happen here until or unless people begin making buying decisions based on security," says O'Gorman, an Offensive Security instructor, penetration-tester, and author of *Metasploit* and *Metasploit: The Penetration Tester's Guide*. "And in the past, the market has demonstrated that it does not make decisions that way. You can look at the home router market as an example of this. Look how many problems have come up over the past few years with a device that is sold partly based on the security functionality that it supposedly provides."

Because there is no single IoT protocol in use (and in fact, very little that's standardized about either the devices or the mechanisms they use to connect to one another), security problems are bound to arise, and incompatibility issues will almost surely surface. It may be that the most effective safeguard, unless and until standards do emerge, is to minimize your IoT connections to the outside.

Jim O'Gorman is a security expert and author of *Metasploit* and *Metasploit: The Penetration Tester's Guide. Image courtesy of Jim O'Gorman.*

"From a consumer standpoint, the best thing that people can do is to not have external network access available for any device that only needs local access," says O'Gorman. "For instance, if I have networked lightbulbs in my house, it would make sense for me to control them when I am attached to the home network. However, there's no real reason to allow the lightbulbs to talk to the Internet."

It's Not *Really* New

There's really nothing new about connected "things." Machines have "spoken" to other machines for many years, first just mechanically (think of clockwork-type gears and cams that cause levers to move, actuating or otherwise affecting the behavior of a second nearby machine), then electromechanically, and then via digital signals. More recently, M2M (machine-to-machine) communications have been the province of large industrial or commercial concerns. Robots in a GM plant, for instance, communicate with parts bins or with supervisors, signaling the need for more paint or bolts, or indicating that it's now safe to move an assembly down the line to the next station.

As one example, consider a commercial printing press. Large printing presses can sprawl for many yards, with one press filling a small building. A computer connected to that press can send information from one assembly to another, and can in turn receive information about ink levels, paper jams, and more from the massive set of systems it's controlling and with which it communicates. When levels run low, the system can signal another device (or a human overseer), requesting replenishment; when machinery falls out of spec, the system can note that tolerances need to be adjusted.

One such system that enables this sort of communications is SCADA: Supervisory Control and Data Acquisition. SCADA gathers and analyzes real-time telemetry data, and uses that information to monitor and control equipment in a variety of industries. SCADA has been around since the 1960s, so there's certainly nothing new about software communicating with remote hardware to monitor and control industrial devices.

But the sheer number of devices now being connected (or about to be connected; the International Data Corporation estimates that the Internet of Things is expanding at a compound annual growth rate of 17.5 percent, at which rate there will be one hundred billion devices on the Internet by 2020), and the fact that these devices are not located in manufacturing plants or warehouses but in homes and farms and schools and automobiles and other consumer settings, heralds a revolution. That revolution is being driven by a three-pronged explosion of silicon, storage, and connectivity.

Processing Power

First, connecting billions of devices and dealing with the data collected and transmitted by those devices requires processing power, which has become—as Moore's Law predicted—ever more powerful, roughly doubling in power every year, even as the cost of such processing power is halved over the same period of time.

Dr. Ryan Kirkbride is a postdoc researcher in plant genetics at the University of Texas at Austin. *Image courtesy of Ryan Kirkbride.*

This geometrical increase in processing power can be seen in any number of industries, of course, not simply in the Internet of Things. Plant molecular biologist Ryan Kirkbride, for instance, talks about processing the human genome and how the speed with which that's possible has increased dramatically, even as the costs of doing so have plummeted.

"As the machines got better," says Dr. Kirkbride, "we were able to get DNA sequences that were about six to eight hundred base-pairs in length. Which is fine if you want to sequence something that's short. Unfortunately, the human genome is three billion base pairs, and these take a while to run. So, using the technology of the day, they started the Human Genome Project in the 1990s, and spent a decade working on deciphering the genetic code of humans. And it was incredibly expensive; billions of dollars were spent in this endeavor, and many, many labs were involved. But all the while, the machines were getting better, and there was a real shift in the mid-2000s, where the bottom essentially fell out of the cost, and that's due to some of the next-generation sequencing technologies that we have. So instead of sequencing a human genome over the course of ten years at a cost of billions of dollars . . . The Holy Grail now is the $1,000 genome—and we're getting close!"

And one reason that scientists are getting close to sequencing a human genome at a cost of only $1,000 is that processing power has improved over the years.

Storage

When we think about the IoT, we're talking about lots of data. Most of it will be in small chunks (daCosta, quoted earlier in this chapter, advocates a very small, very efficient data unit he calls a "chirp"), but we're talking about billions of devices, each producing billions of chunks of data. Some portion of that data (and metadata: data *about* that data) has to be stored someplace, at least temporarily. Luckily, the cost of storage has plummeted to almost nothing.

To illustrate that precipitous drop, consider that a 26MB drive cost some $5,000 in 1980; obviously, only large corporations or government entities could afford that at the time. (Then again, in 1980, who else would need that sort of storage?) Fifteen years later, Seagate was selling 1GB of storage for less than $1,000, and by mid-2009, Hitachi could sell a 1GB drive for about $75. At press time, I can buy a 3TB desktop computer drive (that's three trillion bytes of storage) for less than $140. In other words, storage now costs fractions of a penny per megabyte, and the affordability of storage is one thing that will help make the IoT possible.

Connectivity

Finally, consider the connectivity that underlies the Internet of Things. After all, it is the ability to connect devices to networks—and thus to one another—that makes the IoT workable at all. In order to form a network—any kind of network—one node has to connect with another node and then send and receive data. A few years ago, when all networking was essentially hardwired, the IoT would not have been possible, not at the scale we're envisioning today. Are you really going to dig into your walls to drag a Cat 5 cable from your basement router up to your refrigerator? And then do it again for your thermostat, and again for your hot water heater, and so on? What in the world will you do when you want to connect your toothbrush to the Internet of Things?! (No, seriously. There are now Web-enabled toothbrushes.)

But these days there are a variety of wireless protocols and services that enable devices to communicate with one another without the necessity of dragging wires through walls: From Wi-Fi to Bluetooth to ZigBee to 4G cellular and more, we now have any number of wireless data transfer possibilities, and any or all of them could be used to connect devices to the IoT and to send data to your computer, your smartphone, or your tablet. (All of which, of course, are themselves also connected to the IoT. It may be that we're getting a bit overconnected.)

So, is all of this hyperconnectivity good or bad? Or is it both good *and* bad? That depends partly on your perspective and on how much faith you place in the people who build such devices, and how much you trust those who collect and use the data generated by them. Given the history of such devices, and of data collection in general, it may be wise to view with a grain or two of salt the supposed "empowerment" that results from the sudden ubiquity of IoT-related technologies. But, as with social networking, perhaps we can take heart in the fact that it is in fact ubiquitous and affordable, and that as a result, it may mean that we can use information provided by the IoT against those who might attempt to use it against us. After all, even spies can be spied upon.

The volume and usability of that information is about to skyrocket. Soon we'll be using context-sensitive devices meant to augment reality—from glasses to goggles and from smart watches to smartphones—to expand our reach, enhance our brains, and to make connections that might have been missed before.

But along with those connections come some serious risks, many of which we may not have considered before leaping blindly onto the tech and social networking bandwagons. The final chapter explores what some have called "the death of privacy."

Chapter Twelve

Everyone Is Watching You: The Death of Privacy

Just because you're paranoid doesn't mean they're not after you.
—John Yossarian in Joseph Heller's *Catch-22*

Target Targets You

In February of 2012, *The New York Times Magazine* broke a story by writer Charles Duhigg in which he laid out a frightening tale that purported to describe how illusory our naive notions of privacy might be.

The story, subsequently picked up by *Forbes*, the *Wall Street Journal*, and other outlets, detailed just how much retail giant Target Corporation knows about their customers (that's you, possibly even if you've never set foot in a brick-and-mortar Target store), and how they go about collecting and using that information.

It was a scary (or possibly scaremongering) story, and its centerpiece was an alarming (and perhaps apocryphal) anecdote concerning a young woman who had become pregnant but had not yet found a way to tell her parents. Nonetheless, Target—which had been perfecting ways of mining customers' data in order to make predictive analyses about their buying habits—seemed to know this young lady's secrets. The company began sending the young woman coupons and ads for baby products, children's furniture, and the like.

Needless to say, her father—unaware of the pending new addition to the family—was nonplussed. He charged into the local Target store, demanding to speak to the manager, wanting to know what on earth the store thought it was doing, sending ads for baby wipes, changing tables, and other such products to his impressionable young daughter. Were they trying to *encourage* her to become pregnant, he wanted to know?

Of course, the store manager—not being privy to the data-mining machinations at the company's Minneapolis, Minnesota, headquarters—had no idea what was going on, and he apologized profusely, promising to get to the bottom of the issue.

The manager, who obviously values his patrons and believes in good customer service, even called the man back some days later to apologize *again*.

That was when the father sheepishly noted that, in fact, it was *he* who should be apologizing to the retailer. It had become apparent at home that, unbeknownst to the father, the young lady actually *was* pregnant. Target's data-mining operations and customer tracking *had* in fact yielded on-target predictive analysis data confirming that, based on products the young woman was buying (none of which had anything directly to do with pregnancy or child-rearing, by the way), there was some statistical likelihood that she was pregnant.

Or so the story goes. Parts of the story are certainly true, although the whole bit about the father seems a bit too pat. (One predictive analysis practitioner goes so far as to say that the story has been "essentially debunked," but does not note exactly what he means by "essentially," and has not explained why the "debunking" itself has not become news.)

In any case, it is undeniably true that Target—and most other large retailers—have teams of scientists (including psychologists and statisticians) hard at work trying to find ways to use what we've come to call "Big Data" to connect the dots among their current customers and their potential customers, with an eye toward turning the latter into the former.

The company's goal is not just to sell you *more* stuff; they want to sell you *all* of the stuff. Everything you've been buying elsewhere, Target wants you to buy there. Its goal is to become the source for *everything* you want—often before you even know that you want it. As a Target analyst told the *New York Times* writer, "Just wait. We'll be sending you coupons for things you want before you even know you want them. . . . If you use a credit card or a coupon, or fill out a survey, or mail in a refund, or call the customer help line, or open an e-mail we've sent you or visit our website, we'll record it and link it to your Guest ID. We want to know everything we can."

You Are a Valuable Commodity

Well, of *course* they want to know everything they can. *Everyone*, it seems, wants to know everything they can about you. Why? Well, because they want to sell you things. Or, perhaps a bit more troubling, they want to sell your information *to others* who want to sell you things.

And we're making it very easy for them to collect and analyze that information. We're willing (more than willing, it appears) to sacrifice some measure of privacy, so long as that sharing is perceived as earning access to free services in return: coupons, priority shipping, shopping tips, social networking, information in general. The list goes on, essentially infinitely.

Some 829 million people use Facebook every day. We run about 6 billion Google searches every day. Well over 150 million people use Instagram every month. We send half a billion tweets every day. Every time we do any of that, we share information about ourselves, both knowingly and unknowingly. The former occurs when we post information for friends, acquaintances, or even strangers. The latter occurs when marketers use various means to collect information with an eye toward selling product or by government actors seeking to . . . do whatever it is that such actors do.

Google, wildly successful at presenting to the public a wide variety of "free" services, knows *much* more about you than you might think. (In some ways, the company might even know more about you than you know about yourself.) If you check Google's collection of info about you (go to https://history.google.com/history), you'll find that the company tracks your search history—hour by hour. Google knows the sites you visit and when you visit them, the ads you view, books you've looked at on the Web, and any shopping you might have done that was initiated through Google, or through one of its many partners.

The "partner" issue is a significant one, because these partnerships create intricate webs of connections and information flow among retailers, data aggregators, and marketers. If you use a tool that tracks the trackers, such as Disconnect or Ghostery (discussed in more detail below), you may be surprised to find that, when you visit a website, present on that website are widgets that feed to the site's partners fairly detailed information about your online activities. (In theory, none of this information is "personally identifiable." In practice, experiments have shown that it's not difficult to de-anonymize such data.)

For instance, when you visit CNN.com, you're exposing the visit and your activities on that site not just to CNN, but to a wide variety of CNN's partners. Some of those partners are to be expected, and it's difficult to see how CNN could operate an efficient site without them: Facebook is watching, of course, because there's a "Like" button on the page or a "log in through Facebook" option. Google Analytics (or a similar site) could be watching and helping CNN collect and compile data about which of its pages are visited, in which order, etc. In other words, there are utilities on the CNN site that watch you in order to help CNN build its business. (This is perfectly understandable, though understanding it does not necessarily mean one condones it, and it does not mean that such partnerships are without risk. You tend to think of Facebook, for instance, as a "social network." But that's not why it exists; that's a by-product or a means to an end. It's *really* just another marketing partner.)

You've seen this too: After researching the purchase of a new truck, I began noticing ads and photos such as this one (from Wikipedia's history of the Chevrolet Silverado) popping up no matter where I went on the Internet. Marketers track where you (or more accurately, your computer) have been, and they sell that info to interested parties—such as your local Chevy dealer. *Image licensed under the Creative Commons Attribution-Share Alike 2.0 Generic license.*

But also watching on CNN's site are a variety of partners whose job it is to collect information purely for marketing purposes. (The last time I checked, these included Visual Revenue, Gigya, Livefyre, Nielsen, Outbrain, and others.) The reason CNN and other sites allow these partners to eavesdrop is that, one way or another, there is money to be made in doing so. (This is why when you do a Google search for a wiring harness for a 2002 Ford Taurus, you start seeing ads for Ford wiring harnesses everywhere you go. Your search was tracked, and your interest was noted and shared, so be careful what you search for. This is also why there was such a furor in January of 2015 over the "data leaks" discovered on the Healthcare.gov website. The government site had—some would say foolishly—partnered with several third-party trackers, and the links and other info it was sending to those trackers turned out to contain personal health information.)

Yes, They're Watching

And there are plenty of reasons to be worried. Everyone from your boss to your grocer to your government is either tracking you or buying data from someone who *is* tracking you.

If you carry a smartphone and have not taken steps to restrict their activities, advertisers are most likely tracking you, resulting in targeted ads that can follow you from site to site. (Whether this is a bad thing or a good thing is open to debate, of course. Many of us are fine with that sort of advertising, provided it is both accurate—don't show me ads for things in which I couldn't possibly be interested—and unobtrusive.) There are ways to restrict this activity to some extent; check your user's manual to find out how to opt out of "interest-based" ads.

Of course, advertisers—and others—could also be tracking your *physical* whereabouts, and simply turning off your device's GPS function may not stop them. Interested parties with the correct access can use cell tower–based triangulation to get a fairly accurate reading (say, within a block or so) on your location.

Be aware, though, that research seems to indicate that it's possible, given enough data points (keep the term "big data" in mind, here), to track you down even if you *don't* share your location on a social network. It's more than a little creepy, but you need to know that, given enough background information, researchers are quite successfully predicting your location based on your *friends'* locations.

In fact, the confluence of big data and social networking has occasioned several worrisome privacy-related developments. Raytheon, for example, is working on a program called RIOT: Rapid Information Overlay Technology. Its purpose is similar to some of the research cited above. RIOT sifts through multiple social media sites, seeking to track a person of interest. The system displays a web-like visual representation of the target's interactions, and has in some cases been able to accurately predict the person's next move.

When you combine data leaks (whether accidental or purposeful) with big-data analytics, you end up with a privacy nightmare.

Even your car is spying on you.

Your car's "black box," or electronic data recorder (EDR)—you did know that most late-model cars, and all cars sold in the United States after 2014, have one of those, right?—has long been collecting basic data on your driving habits, on seat-belt status, and on conditions just prior to and after any sort of accident. (The first ones appeared back in 1974, as part of systems used to help deploy airbags in the event of a collision. They have since become much more complex and much more ubiquitous.)

As EDRs become more sophisticated and more prevalent, some are beginning to worry about potential privacy issues associated with them. A recent Future of Privacy paper notes that, "In the future, in-car technologies will increasingly gather information about driver behavior or their biometric data. For example, vehicles will be able to quickly identify their drivers, changing car settings to accommodate the driving profile of a teenage or elderly driver. Sensors in the steering wheel or driver's seat will monitor stress-levels and health conditions. Much of this information is used to drive vehicle safety improvements."

The data has indeed been used to help manufacturers build safer vehicles, but it has also been used in court, and may at some point be used by insurance companies seeking to either assign blame or apportion premiums based on your driving habits. (Note that data from EDRs has been used in court both to convict *and* to exonerate.) It's not difficult to imagine a time when your car simply refuses to start, because it has sensed that your stress (or alcohol) levels are too high. Or perhaps it simply knows that you're behind on your payments.

And then there's Flo, the perky, ever-helpful sales agent from those Progressive Auto Insurance commercials. Flo may be cute, but she turns out to be kind of nosy: If you opt in to Progressive's Snapshot program, the company will begin collecting data about your driving. The Snapshot device plugs into your car's diagnostic port and collects data on your driving habits. At the end of every trip, the collected data is sent to Progressive via a nationwide cellular network. Progressive then uses algorithms to convert the speed readings into info about mileage and hard braking events. If the data shows that you're a mature, careful driver, you could qualify for reduced premiums or other benefits. (Currently, Progressive collects only time and speed data, but the company is looking into adding GPS and other info to that data stream.)

Is such data collection a bad thing? In many ways, no. After all, why *wouldn't* an insurance company want to make sure that its customers are safe drivers, and why wouldn't it want to reward them for their prudence? And wouldn't you want to *be* rewarded for being a good driver? It's not that this sort of data collection is inherently bad; it's that it's another link in an ever-lengthening chain of data collection and possible abuse, yet another potential security leak, and yet another data stream that contains your information, which could one day be used by unknown actors for as yet unknown purposes.

In and of itself, it might be a perfectly acceptable way to collect and use information about my behavior, but it begs other questions: Perhaps I trust Progressive; after all, we love Flo. But what about other people with whom Progressive might share the data? Do I trust *them*? (The data has moved from one trust context to another, most likely without my even being aware of it. See "Privacy and Context," below.) Is the data safe? Secured? Protected? What if it's stolen or hijacked? Who controls it then? In what ways might it then be used against me? Who *owns* this data, anyway? Me? Progressive? Some partner? And on and on. Information is transitory and fluid; once it's out of your hands, it's very difficult to control its use or to know where it might end up.

Your car is collecting data in other ways, too. With the rise of "the connected car," it's now becoming common for your vehicle to connect to the Internet, and it's about to become a lot more common. When that happens, your car becomes a rich source of information, not only for the manufacturer, but also for organizations with which the manufacturer partners: insurance providers, the government, marketing partners, and others. (Not to mention hackers, who will likely try to find ways to break in even *without* an invitation from the vehicle manufacturer.)

Consider Ford Sync. This is a sophisticated system that can be very useful to drivers. It includes GPS navigation, voice recognition, text messaging, and mobile apps, as well as integration with Apple's Siri and with a number of music services. Handy stuff.

But as one Electronic Frontier Foundation attorney noted, it's almost impossible to know with whom data collected by such systems is being shared. Sync's terms of service note that if you use the system to dictate e-mails, it will collect location and call data. Ford then says it will share that data with "business partners" in order "to improve service," according to the EFF. Both statements are so vague that the car owner is left woefully uninformed—even assuming that she reads the terms of service, which is doubt-

Ford's Sync is a handy tool that offers various forms of connectivity, including access to music services. It also collects data on your vehicular activities, and Ford may share that data with its partners. *Image courtesy of Ford Motor Company.*

ful to begin with. The owner will have no idea *what* data is being collected, when, how, or how often. And she will certainly have no idea who Ford's partners are or what use *they'll* make of the data collected by the system.

Another way to collect driver data? In-car cameras.

Hertz began putting cameras in its rental cars as part of its NeverLost navigational device in 2014. The idea seemed straightforward—and innocent—enough: The camera feature was added so that the company could, at some point in the future, provide a service that allowed it to connect a customer in a car to a live agent via video. This could have proven helpful in the event of an accident, a mechanical problem, or a question. However, neither the service nor the camera was ever activated, according to the company. "The camera feature has not been launched, cannot be operated, and we have no current plans to do so," said a Hertz spokesperson.

But let's face it, it's creepy. The company says that if it ever did decide to use it, it would be on an opt-in basis, but how do we *know* that? Could someone hack into the camera and turn it on? Could a Hertz employee, for instance, tap into it without company supervisors knowing?

In any event, it turns out that in many states it's illegal to videotape someone in your vehicle without them knowing it. The Chevrolet Corvette's "nanny cam" allowed paranoid owners to monitor parking valets, but the company had to immediately warn new car owners not to use the feature because it could be illegal to spy on people without them knowing they were being recorded.

Government Actors

Any of us who might have previously been operating under the naive assumption that government agencies would be scrupulous about obeying laws relating to the collection and retention of private data have recently been disabused of such credulous notions. With the Edward Snowden revelations, it's become obvious that governments (sometimes for good, or at least, understandable reasons) are not above bending (or occasionally ignoring) the law in their quest to capture criminals, promote the general welfare, and protect us from the threat of terrorism. (And we do want to be protected, after all; that is one of the reasons we pay to have a government. But it brings to mind Ian Bell's question in a recent Edinburgh, Scotland, *Herald* editorial: "Why do the governments of the West, endowed with so much military, economic and scientific power, always need a few more of your freedoms to win their battles in freedom's name?")

Case in point: The US Marshals Service is a storied and respected federal agency, part of the Department of Justice, long tasked with fugitive apprehension and transport, witness security, and the protection of federal judges, among other things. Seemingly out of character, in 2007, the Service began operating clandestine flights out of several metropolitan airports.

The Marshals Service flew (and, as far as we know, still flies) small private aircraft, usually Cessnas, and on board each aircraft was what's known as a "Stingray" device—a cell-tower emulator. It fools cell phones into connecting to *it* rather than to a legitimate tower. The signal is eventually passed to an actual tower and the call is completed normally, but with the user unaware that a covert device has collected data from the call before passing it on. Obviously, such tools are useful in combating crime and terrorism, but privacy advocates note that there are grave concerns regarding the use of such tools.

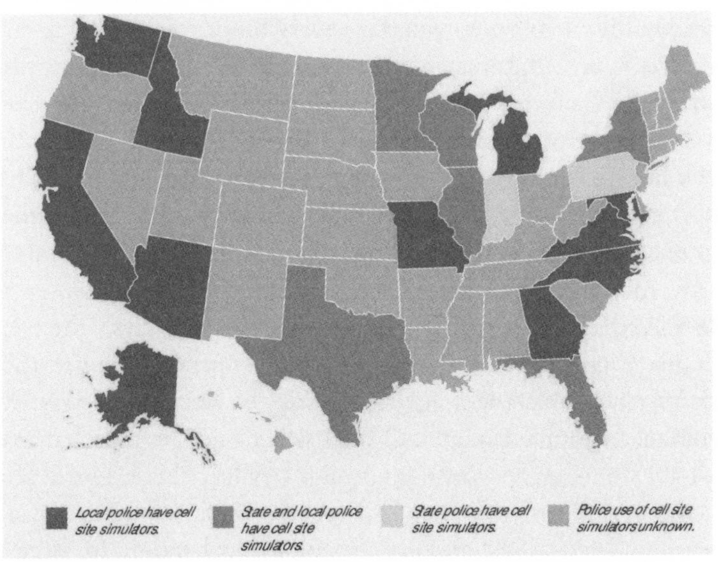

According to the ACLU, many federal agencies utilize cell-tower emulators—also known as "Stingrays." These include the FBI, DEA, US Secret Service, the NSA, and the US Marshals Service, along with almost every branch of the US Armed Forces. Stingrays intercept cellular calls, capture the data from the call, and then pass the call on to a legitimate (i.e., real) cell tower. *Image by Lori Garris.*

"This is a disturbing progression of the federal government's use of this technology," said an ACLU attorney in a CNNMoney interview. "What's different about this . . . is that it vastly increases the number of completely innocent bystanders whose information is being swept up by law enforcement."

Then again, as the Snowden revelations have shown, collecting random cell call data is nothing compared to what some agencies have been able to do.

Consider the NSA's MYSTIC program. Begun in 2009 and fully active as of 2011, MYSTIC is a "retrospective retrieval" tool that can replay actual voice calls from any call that took place within a given country. Yes, that means that the NSA is presumed to be able to capture and store *every second of every telephone call made in that country.* MYSTIC relies on information retrieval that's predicated on the ability to collect 100 percent of a country's telephone calls.

Is MYSTIC a good thing or a bad thing? Well, one assumes that context matters. If the NSA is recording every bit of every telephone call that takes place in, say, Afghanistan or North Korea, many Westerners would be fine with that. But what if the MYSTIC program targeted, say, Israel? That could be a more complicated scenario. What if it were aimed at France? What if it were operating within the United States itself? Then what?

(Of course, NSA or other government overreach is also a problem in that it encourages *other* nations—not all of them democracies in any sense of the word—to use our government's overreach to justify their own. And as the Electronic Freedom Foundation's Jillian York says, "the number of governments using the NSA's surveillance to justify their own is a huge threat. What the NSA is doing is horrible . . . but for the majority of us, the consequences are minimal. On the other hand, when China or Saudi Arabia implements the same type of surveillance, the consequences are severe: prison or death." And, as York also points out, we don't have good information on the Middle Eastern countries with which the United States is sharing data and how it's being used by those countries' security services.)

Once the security genie is out of the bottle, it's difficult to put it back. And it is the nature of covert operations, however well-intended, to elude the eyes of those charged with ensuring that such technologies are not misused.

What, me worry? Sure, I'm paranoid; but am I paranoid enough?

Apparently we do worry about the whole privacy problem, but perhaps not as much as we ought to.

A recent Pew survey found, for instance, that 91 percent of adults surveyed agreed that consumers have lost control over how personal information is collected and used by companies. Meanwhile, 80 percent of those who use social networking sites say they are concerned about third parties like advertisers or businesses accessing the data they share on these sites, and 70 percent of social networking site users say that they are at least somewhat concerned about the *government* accessing without their knowledge some of the information they share on social networking sites.

We worry about it, but we don't seem to *do* much about it.

Year after year, most of us use weak passwords, tempting fate with such easy-to-guess "secret" codes as "password," "123456," and "qwerty." (And for some reason, "monkey," which always lands in the list of the top worst—but commonly used—passwords. Surely there's a sensible explanation for that.) Not only that, but we use the same weak passwords on multiple sites, in spite of almost daily admonitions from security experts not to do so.

At the same time, we leave gaping holes in our social networking settings, sharing just about everything with friends, bare acquaintances, and sometimes with complete strangers. As security expert and author Chris Hadnagy comments, "I can probably profile 90 percent of the people that use [the three major] social media outlets profusely. And if you don't have them locked down and they're public, then you are an open book to a social engineer."

Another Pew poll (released about two years after some fairly startling revelations about the US government's proclivities for digital surveillance) shows that a vast majority of Americans recognize that their digital lives aren't secret. But at the same time, a clear majority of respondents said they weren't overly concerned about the government snooping around their calls and e-mails.

As a whole, we're apparently a pretty complacent lot, and we all know that young people are the worst, seemingly not worried at all about privacy. There is certainly the perception that young people are blissfully unaware of the erosion of their privacy—or perhaps that they are well aware of it, but don't seem to care.

There is something to both those perceptions, but it may also be that we're looking at it simplistically.

"The idea that teens share too much—and therefore don't care about privacy—is now so entrenched in public discourse that research showing that teens *do* desire privacy and work to get it is often ignored by the media," says author and researcher danah boyd. (For personal and political reasons, Dr. boyd does not capitalize her name.)

In *It's Complicated: The Social Lives of Networked Teens*, Dr. boyd notes that oversimplifying teens' privacy concerns is a mistake.

"The teens that I met genuinely care about their privacy," she says, "but how they understand and enact it may not immediately resonate or appear logical to adults. When teens—and, for that matter, most adults—seek privacy, they do so in relation to those who hold power over them. Unlike privacy advocates and more politically conscious adults, teens aren't typically concerned with governments and corporations. Instead, they're trying to

Dr. danah boyd is the author of *It's Complicated: The Social Lives of Networked Teens*. She argues that teens *do* care about privacy, but that they view it differently than many adults. *Image courtesy of danah boyd.*

avoid surveillance from parents, teachers, and other immediate authority figures in their lives."

The popular social media sites, says boyd, are designed to encourage participants to spread information, rather than to contain it. Thus, it's far easier to share something with all friends on Facebook than to limit the visibility of a particular piece of content to a narrower audience.

"As a result," says boyd, "many participants make a different calculation than the one they would make in an unmediated situation. Rather than asking themselves if the information to be shared is significant enough to be broadly publicized, they question whether it is intimate enough to require special protection. In other words, when participating in networked publics, many participants embrace a widespread public-by-default, private-through-effort mentality."

Teens do think through the social cost of what they post, but they don't always get it right, says boyd. Still, she's impressed by how teens utilize a fairly sophisticated technique she calls "social steganography" in their online interactions.

Steganography is a cryptographic method in which data is hidden in plain sight. Often, this means that information is encoded in, say, an image. If you know to look—and if you know exactly *where* to look—in the image, or in the raw data that makes up the image, you can view a message that was hidden in the picture.

Social steganography, says boyd, is the practice of "hiding messages in plain sight by leveraging shared knowledge and cues embedded in particular social contexts."

In other words, young people use linguistic and cultural tools—lyrics, in-jokes, cultural references, and more—to encode messages that appear to be about one thing (or possibly about nothing) when in fact they're about something of which casual viewers (read: parents, teachers, enemies, gossips) are unaware.

"Whole conversations about school gossip, crushes, and annoying teachers go unnoticed as teens host conversations that are rendered meaningless to outside observers," she says.

Many young people perform entire intricate and protracted social dramas online, but a parent (or out-of-the-loop friend) would have no idea that the drama is even taking place.

The upshot is that young people apparently do have enough sense to want to ensure that there *is* privacy within certain contexts—and they're smart enough to recognize and distinguish among those various contexts. Perhaps the problem is that they're not very good at anticipating *future* contexts, many of which might turn out to matter very much in terms of the privacy they've given up previously, possibly without even recognizing that they gave it up.

Privacy as a Cultural Value

Some researchers feel that US residents do not feel about privacy the way much of Europe does. Other countries take their privacy much more seriously, says Jillian York, the Electronic Frontier Foundation's director for International Freedom of Expression. Based in Berlin, York believes that we have a serious problem, but that it tends to get overlooked, especially in the United States. We simply don't care enough, she says.

"I believe we have a looming privacy problem," notes York, "but it's on us to change that: It doesn't have to mean banning tools or concepts. Rather,

Jillian York is the Electronic Frontier Foundation's director for International Freedom of Expression. Based in Berlin, York notes that US citizens don't seem to care enough about the erosion of privacy. *Image courtesy of Jillian York.*

we as a society need to decide what privacy means to us and in what ways it's important."

In the United States, privacy simply doesn't have the same weight as it does elsewhere.

"An interesting analogy here," says York, "oddly enough, is public toilets. In the United States, the doors to toilet stalls come only to one's shins, whereas in Europe, they are from floor to ceiling. In other cultures, they may be barely present or even nonexistent. Privacy is a cultural phenomenon, and in the United States it simply doesn't have much value."

We really need to learn to value privacy more in the United States, says York.

"In Germany, privacy is important because of the country's history," she says. "People understand the risks of surveillance because they've experienced them. The majority in the United States, or at least the chattering classes, has not, and therefore does not see the potential harm."

York doesn't blame the technology itself for the problems. Asked about privacy issues related to social media, she notes, "Social media certainly poses risks to our privacy, but it's the corporations behind the sites, what they're allowed to do, and what they *desire* to do with our data, that is the real problem.

The business model, not the technology itself, is the issue, says York.

For example, she says, "Facebook requires that users create accounts using their 'real' names, or the names that match what's on their identification. There's no legal reason for this, nor has it been proven to promote 'civility,' as the company might argue. The (real?) reason becomes clearer, however, when you think about the fact that user data becomes far more valuable when it's attached to the name of a living, breathing, identifiable person."

What she means, of course, is that the data is much more valuable when attached to the name of a living, breathing, identifiable person *to whom products can be sold* and whose information can be brokered, sold to still more marketers.

As privacy and software freedom activist Richard Stallman has said, "Facebook is not your friend. Facebook is a surveillance engine."

On the . . . upside?

Watching the Watchers

Previous chapters have included "On the downside . . ." sidebars in which, while noting the often-positive results of the technology being discussed, some of the *negative* aspects of the technology were also pointed out. That's difficult with this chapter, because most people would view the erosion of privacy—the subject of this chapter—as entirely negative already.

So perhaps it's time for an "On the upside . . ." sidebar. There is, after all, a positive aspect to the whole privacy debacle, and it boils down to the fact that the very technologies that can be used against you can also be used *by* you to protect yourself.

Cameras, for instance. Ubiquitous, powerful, and increasingly hard to spot, cameras can be used against you, often without your knowledge. When you zoom past a "red-light camera" at an intersection, do you even notice it? If a small drone flew a few hundred feet overhead, its camera's lens pointed at you or your friends partying at the local park, would it attract your attention? If a government agency wanted to surreptitiously acquire photos of you engaged in some activity (legal or not), would it be able to do so? If local or regional law enforcement were to scan your license plate, would you even know that you'd just become one of the billions of plates being stored in (mostly private) databases?

Chances are that in all of those cases, you or your data could be recorded. But the same cameras that law enforcement—as one example—can point at you, can be pointed (and *have* been pointed) back at them. In fact, given the ubiquity of smartphone cameras these days, it's an unusual public arrest or officer-mediated altercation that goes unrecorded by citizen video.

And, by and large, that's a good thing; it means that the police can more easily be held accountable for their behavior. (Keep in mind, though, that a brief video clip does not provide a full and balanced record of what occurred. The police officer's job is a difficult and volatile one, charged with political nuance and fraught with dangers of all sorts. It's unfair to pretend that one video snippet, perhaps shot by a biased observer, will tell the whole story of what is likely a complicated and perhaps extended incident. But some record is likely better than none, and the fact that non-officer video is available to contradict—or corroborate—the officers' version of a story surely must lead to a more-accurate telling of that story.)

In the same vein, police bodycams, increasingly used by law enforcement agencies around the world, would tend to have the same effect, since such recordings can help to corroborate or contradict the officers' version of an incident. (The cameras can create privacy issues of their own, of course, if recordings are made in private dwellings or of protected actions. But, as Helen Nissenbaum notes in *Privacy in Context*, at least with such cameras, "individuals are equipped to monitor and

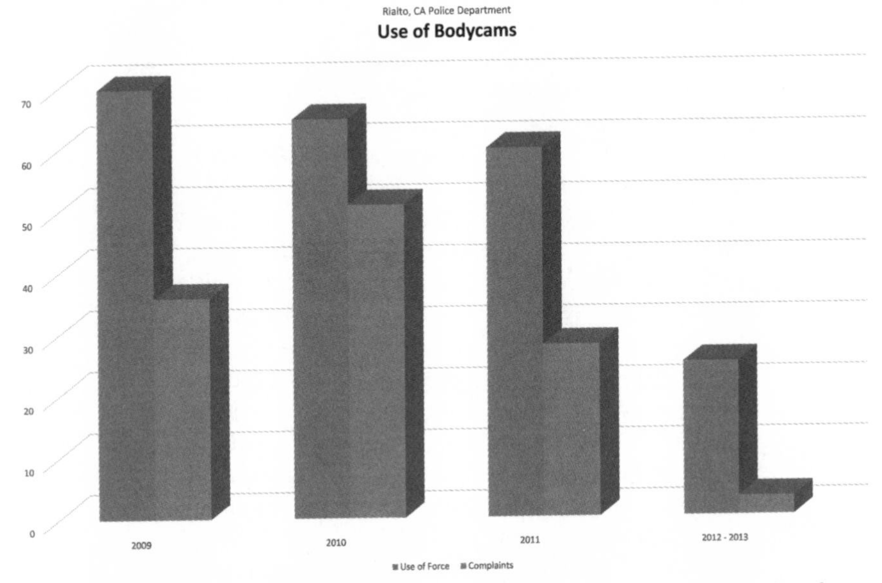

The Rialto, California, police department (a midsize department in a community of about 100,000 residents) began an experiment in 2012 in which officers began wearing bodycams. Use-of-force incidents plummeted, as did citizen complaints. *Author image; data courtesy of Rialto, California Police Department.*

track one another as well as authorities, offering a glimmer of hope at a more level playing field while fueling the worry that watchful eyes are now inescapable.")

Interestingly, the ACLU, while it generally argues *against* the proliferation of surveillance cameras, notes that the benefits of police usage of on-body cameras can act as a check against the abuse of power by officers, and can serve to protect both the public and the officers. The group's policy statement notes:

> *Historically, there was no documentary evidence of most encounters between police officers and the public, and due to the volatile nature of those encounters, this often resulted in radically divergent accounts of incidents. Cameras have the potential to be a win-win, helping protect the public against police misconduct, and at the same time helping protect police against false accusations of abuse.*

This position results in a certain irony: The ACLU, on record as being against video surveillance of public places, is at the same time *in favor* of video cameras in police cars and prisons, and also of the filming of interrogations. The group's conclusion is that, if the right policies are in place to protect both the public and the officers, police bodycams are "a win for all."

But cameras need not be the only tools in your arsenal. Previously mentioned tools such as Disconnect and Ghostery can literally "watch the watchers" for you.

When you visit a website and take its measure, there are unseen actors taking *your* measure. That is, when you visit CNN.com, FoxNews.com, Dictionary.com, or Chevrolet .com (but not, interestingly enough, CIA.gov—as far as we can tell, anyway), there are other agents watching your interactions with the website. Some of them (such as Google Analytics) are simply there to collect usage info for the site owner, but many of them are marketing organizations, or are affiliated with such organizations.

For instance, on the Chevy.com site, one watcher (of well over a dozen) is BrightTag, run by Signal. (BrightTag is now simply known as Signal, after the company that created it.) Signal says that its mission is "building a new data infrastructure for the future of marketing technology." On its home page, it says, somewhat more plainly, that the company "connects your customer data into a single identity" and that it helps its customers "collect 100% of your engagement data." Signal is, says the company, "the global leader in cross-channel marketing technology."

Are these bad things? Not necessarily. An awful lot depends on what Signal does with that data, how scrupulously it's guarded, how well it's anonymized. If Signal's scrutiny results in you getting information that you seek, or if it helps you to find products that you need, then the company is ultimately helping the advertisers help you.

But how do you *know* how careful they are with the data? Can you make an informed decision if you don't even know the data is being collected in the first place?

You can't, which is why tools such as Ghostery (https://www.ghostery.com/en/) and Disconnect (https://disconnect.me/), among others, are important. These programs allow you to watch the watchers. If you feel that too many companies are watching—or that their watching is inappropriate—such utilities let you block any or all of the watchers. When you click that Block button, you've just used the Internet against the people who are using the Internet to spy on you.

Privacy and Context

Dr. Celeste Campos-Castillo, a postdoctoral research fellow at Dartmouth College and assistant professor of sociology at the University of Wisconsin, points out that *context* is key to discussions about privacy.

"When we share information in a certain context, we (implicitly or explicitly) expect it to *stay* in that context," she says. "For example, I share sensitive pieces of information with my doctor, and care very little if the doctor shares it with my insurer. I may, however, care if it leaves the health care context and enters into the realm of scientific research or media, in the case where my disease becomes national press—think measles outbreaks."

Thus, the fact that we are outraged over privacy breaches fits in with the idea that trust in another has much to do with context. I may trust you

Dr. Celeste Campos-Castillo is an assistant professor of sociology at the University of Wisconsin–Milwaukee. She specializes in communication and information technologies and social psychology, and argues that context matters very much in privacy issues, or in any technology-mediated situations in which trust must play a part in the interaction. *Image courtesy of Celeste Campos-Castillo.*

to drive my car, but perhaps not to cook me dinner, says Campos-Castillo. In fact, if we're just getting to know one another, I don't even *know* how context-specific my trust is until we enter some new context, perhaps going from car driving to cooking.

"In other words," says Dr. Campos-Castillo, "something needs to change to trigger this realization."

Helen Nissenbaum, in *Privacy in Context* (a book to which Campos-Castillo often refers during our conversation), characterizes security breaches as instances where information moves across contexts that we are not okay with, and to which we have not agreed. We share our lives with Facebook, she would say, not because we trust them, but because no major breach has occurred that affected us (e.g., no malfunction caused all my photos to be broadcast to those beyond my friendship circle). But once a Facebook breach does occur, *then* I may consider trust versus distrust issues.

Nissenbaum also notes that privacy issues are exacerbated, not caused, by technology. We could always request copies of vital records (births, marriages, deaths) of almost anyone, just by walking up to a brick-and-mortar building and perhaps paying some sort of records fee. No one seemed to care much about that.

All of a sudden, though, "these vital records became accessible online," says Campos-Castillo. "Now there's an uproar. Yet, really, the rules of the game haven't changed. Anyone can still access these records, but the mode of asking has changed."

So, privacy issues are not always *caused* by technology—but they may be exacerbated by it. Information was always available, but few cared, because there were barriers to its access: You needed time, skill, proximity, money, or all four in order to get at it. Private investigators in those days specialized in locating such sources of information (county records, city libraries, motor vehicle departments, police records, etc.), and in cultivating relationships with people who had access to those sources. They still do, but now just about anyone can very easily acquire information that used to take a good deal of effort to locate. The information has always been there and been accessible, but now it's much easier to find, correlate, aggregate, mine, analyze, etc. This has greatly affected our privacy, but it's an effect of *degree*, not of kind.

Dr. Jason Hong agrees that the issue is really one of scale. Hong, of Carnegie-Mellon University, heads up the university's Computer Human Interaction Mobility Privacy Security (CHIMPS) group.

"What's changed," says Hong, "is the scale of information that can now be collected, stored, searched, processed, and shared. Before computers, it was impractical to do all of these. Nowadays, we have both virtual and physical sensors for collecting data, ranging from clickstream patterns to smartphone cameras to GPS. We also now have large-scale hard drives and databases for storing and searching data. Furthermore, there are many kinds of analytics, visualization tools, machine-learning algorithms, and data-mining techniques for finding patterns in the data. Sharing data is also easier than ever, with fast networks, large data warehouses, social media sites, and e-mail. It's not any single technology, but rather the confluence of many different technologies that both makes our lives easier but which also poses new kinds of privacy challenges."

Privacy: A Relatively New Concern

Keep in mind that concern about privacy is a relatively new thing. Certainly we can point to historical periods in which privacy was more or less unheard of. In the Middle Ages, for instance, privacy was not something with which many were concerned.

Take sex, for instance. During the Middle Ages, people—often many people—shared rooms. Children would have been exposed at an early age to the idea of two people having sex on a bed surrounded by adults, children, hangers-on, and—in wealthier households—servants.

As sociologist Svend Riemer noted, "The display of the naked body was not frowned upon. Sleeping quarters were not isolated but were readily shared by all members of the family as well as their servants and their guests. Beds were shared at all ages by non-married members of the two sexes."

The same lack of prudery surrounded the issue (so to speak) of bodily functions. Despite their name, privies were not necessarily private affairs; people were accustomed to seeing to such needs more or less publicly. (Though, for the sake of hygiene, often at a bathhouse removed from their living quarters.)

Neither of these things—or many other such privacy questions—bothered people. As Riemer notes, "this lack of privacy did not cause much suffering or frustration, since the desire for it was highly undeveloped."

Even today, not everyone feels that privacy is an inviolable right, or that an erosion of privacy is necessarily a bad thing. And most would admit that there are times when law enforcement or government has the right to "invade" one's privacy. (Of course, by signing on to Google, Facebook, and the rest, you've *already* given away a certain amount of privacy; why complain when law enforcement takes a bit more of it, especially if it's for a good cause, goes the argument.)

Alexander Nazaryan, writing in *Newsweek*, notes that while it's true that a complete lack of privacy would be harmful, there are sometimes good arguments for government agencies' invasions of your privacy. "Invading privacy to maintain the public welfare is the government holding up its end of the social contract, ensuring that the privacy of a few is not allowed to compromise the life and liberty of many," he says.

Nazaryan also notes that the anonymity provided by maintaining strict privacy can lead to harmful trolling—and outright attacks. "Let's say a female academic is harassed by Twitter trolls who threaten to rape and kill her [remember Gamergate?]," he says. "Twitter may have no legal obligation to reveal the trolls' identities, but doesn't it have a moral one?" (See chapter 6 for more on Gamergate.)

Nazaryan's point is that there needs to be a balance between complete privacy and the information that government requires to hold up its end of the "governing" bargain. Similarly, he seeks to mitigate the issue of online attacks by requiring people to reveal their identities; if you wouldn't say it in a signed letter (or to someone's face), then perhaps you shouldn't be saying it at all, he argues.

The devil is in the details, of course, and in what are bound to be varying interpretations of those details. Who decides how complete the privacy needs to be and how much access to grant law enforcement and government? And, assuming that we can agree on the perfect balance, how do we ensure that government agencies don't overreach and tip that balance?

It's good that Nazaryan isn't too worried about privacy, though, because the website (Newsweek.com) on which he posted his comments is currently transmitting information about site visitors to dozens of activity trackers, most of them marketers.

Promise and Peril

The Internet (and this goes for technologies built on it, such as social networking) is not a safe place. It's not a private place. It's not a secure place. It was built not to *protect* information, but to make it easy to *share* information. Everything we've ever done to attempt to make Internet spaces safe, secure, and private has essentially been a Band-Aid used in a (mostly futile) attempt to cover a gaping wound. We can layer as many Band-Aids over the wound as we want, but it's still just emergency first aid—certainly better than nothing, but not nearly as good as having avoided injury in the first place.

Can we avoid a loss of privacy? Dr. Hong thinks not, and he unknowingly echoes Nazaryan in his comments. "It's not possible in modern society to avoid a loss of privacy, and in some ways, that's actually a good thing. For example, there is lots of information that a government needs to know about its citizens to govern effectively. These might include the census, traffic, and commute information, income levels, health information, and more.

"At the same time, there are a lot of growing challenges to privacy, primarily because it is cheaper to gather, store, search, process, and share information than ever before. The challenges come from all fronts, ranging from overprotective parents and nosy little sisters, to overzealous marketers who try to collect reams of data about people so as to create more-targeted ads, to services that might discriminate against you based on inferred demographics, to governments that might intrude on civil liberties."

The biggest threat, says Hong, is machine-learning techniques for processing the massive quantities of data being collected. The advent of networked computers made it possible to cheaply collect data at a scale previously unimaginable. "What has changed and is still changing," says Dr. Hong, "is our ability to process and understand what that data means."

As of this writing, the latest players to reach for a piece of the data pie (the pie we apparently willingly share with them) are Uber and Radio Shack.

Uber, the company that has very quickly redefined and streamlined the transport of private parties from one place to another, collects data. A *lot* of data, and much of it quite meaningful. Consider that Uber doesn't simply know your name and contact info, which of course it would need in order to arrange your transaction and pick you up and drop you off. The company also knows *where you go*. And when you go there. And how often. And, quite possibly, whether you spend the night. Especially when collected over time, these are important data points; Uber can figure out an awful lot about you by knowing where you go, when, with whom, and for how long.

And other companies are beginning to recognize the value of such knowledge. Uber recently signed a deal with Starwood Hotels that gives guests a chance to earn points when they ride with Uber. The points can be redeemed for a variety of perks, including air miles and upgrades—so long as the guest agrees to link his Uber and Starwood accounts, thus providing both parties with a wealth of information about that particular guest.

As Naked Security writer Lisa Vaas notes, "Given the mountain of data [Uber is] sitting on, it's well on its way to becoming a Big Data company on par with the likes of Google, Facebook or Visa."

And then there's Radio Shack, which declared bankruptcy in February of 2015. As is customary, the company listed its assets, hoping to sell them and thus either satisfy its creditors, emerge from bankruptcy, or make a

To request a ride via Uber, just use the app. The company says you'll usually have a ride ("to match your style and budget") within minutes. *Image courtesy of Uber.*

possible sale more attractive. This time, though, one of the assets listed was customer data—*your* data, assuming you've ever stopped in for a battery, a weather radio, or a length of speaker wire. The data includes the phone numbers, e-mail addresses, and shopping habits of more than one hundred million customers. This is valuable information, and it's all about you. (It amounts to more than thirteen million e-mail addresses and sixty-five million customer names and physical address files.)

In cases such as this, the privacy angle gets complicated. Perhaps back in 2010 (which may have been the last time you—or anyone else—shopped at Radio Shack), you were fine with the company having this information about you. But you couldn't know that, some years down the road, Radio Shack would be looking to sell that data to a third party. And when you signed up for Uber, you thought you were just getting transportation. Instead, it's quite possible that you were—literally and figuratively—being taken for a ride.

In the end, the Internet is about communication—mostly free and largely unfettered communication. But communication technologies *leak*. If you're going to stand on the corner and shout at a person walking by, you need to realize that *other* people in the vicinity will hear you. If we choose to use these tools, we need to do so realizing that there are costs, and that one of those costs is a potential loss of privacy. Is that terrible? Maybe not, at least in many cases. But we need to make informed choices about the tools we use and about the amount of privacy we're prepared to give up. And, if we intend to preserve as much privacy as possible, then we need to learn how to use the privacy tools given to us (e.g., Facebook's admittedly labyrinthine privacy settings) and to *demand* both better tools and more oversight.

And Finally, a Caveat

Privacy is crucially important, possibly the most important topic discussed in this book, and there is absolutely no way in the world to do it justice in one chapter. Like almost every other topic in the book, privacy is important enough—and the ongoing erosion of privacy disquieting enough—to deserve its own book. (And, of course, there *have* been many books written on the subject. Check out the books and articles listed in the bibliography. Read anything—and everything—written by such experts as Bruce Schneier, Chris Hadnagy, or Brian Krebs. They're not only useful and informative publications, but they're written by good writers who make complex, technology-related security issues interesting and easy to understand.)

Keep in mind that it's perfectly okay to decide to give up some portion of your privacy. These days, that's part of the trade-off we make in return for "free" access to social networks, to communicative tools, and to information in general. The key is to be making *intelligent and informed* decisions about the privacy you're willing to give up, and the circumstances under which you're willing to do so. You can't make informed decisions when you don't even know that your privacy is being invaded.

Afterword: The Death of Ötzi

He was murdered on a spring evening, early enough that the sun had not yet set.

The solitary hunter, who throughout the day had busied himself trapping and killing small mammals for their meat and their pelts, gathered his cloak and furs about him and struck out for the foothills. It was rocky and cold, and the snow-brushed stones stabbed at his feet, even through the grass-stuffed leather boots he wore. When he gained the summit, he stopped to look around. He hadn't eaten since morning, and he was looking forward to an evening meal on a nearby outcropping, sheltered from the chilling wind.

Unshouldering his pelts, he dug around inside his bags and found the makings of a decent supper: some cooked wild goat, a large chunk of unleavened bread, some herbs. By his standards, an ample meal, and one he ate in a leisurely fashion, chewing slowly and enjoying each bite.

Stretching, he rose and turned his back to the rocks; he peered into the valley, alert for danger. He had the high ground and felt relatively safe from animal attack or from ambush by other men.

But he wasn't safe at all. The arrow that killed him was launched by a longbow of steamed and bent yew, strung with a plant-fiber bowstring. The arrow was thatched with feathers and gut, and bore a flaked flint arrowhead with razor-sharp edges. It struck him in the back of the left shoulder, the impact spinning him around so that he could gape at his attacker. The arrowhead had pierced an artery, and he fell to the ground within seconds of being struck. In seconds more, he was unconscious; within a minute, his attacker had finished him off with a brutal blow to the head. In the lonely Alps, the dead man lay sprawled on the rocks, blood seeping into the snow.

He lay there for five thousand years, until his frozen body was discovered in 1991 by hikers.

This is the first murder we know about, the earliest killing of which we have direct evidence. It's the story of the mummified Bronze Age body that became known as the Iceman, or, more colloquially, as "Oetzi" (or "Ötzi"), after the mountainous region in which he was found.

The location of the find occasioned an occasionally bitter intergovernmental spat over which country had the rights to Oetzi. Originally discovered by Helmut and Erika Simon, two German alpine hikers, the find was first investigated by the Austrians. Then began a drawn-out argument over exactly in which Alps the body lay—the Italian or the Austrian. In the end, a survey determined that Oetzi's remains were found on the Italian side of the border, but only by about three hundred feet. (Of course, our world being the litigious place it is, that wasn't the end of it. The original discoverers felt entitled to a "finder's fee," as did various hangers-on who appeared later. The case wended its way through the courts for several years. Finally, in 2008 the South Tyrolean provincial government awarded Mrs. Simon—Mr. Simon had died in the interim, his body presumably *not* preserved in snow and ice—a settlement equivalent to about $200,000.)

So, even some five thousand years after he was murdered, Oetzi was still the victim (or perhaps the cause) of a certain amount of rancor and bitterness.

What's missing? Plenty.

But why shouldn't that be the case, really? We are, after all, talking here about the fruits of a technology designed solely to injure or kill. How could it *not* cause rancor and bitterness?

Technology has been killing people since the first club was fashioned and the first spear thrown. The first time an angry ancestor picked up and hurled a rock, the military-industrial complex was born. And ever since then, war and conflict have played a large part in the development of many of our technologies. From gunpowder to GPS, from catapults to computers, and from robots to radar, there are few areas in which military use (or potential use) has not played a formative role in the development and advancement of the technologies of today and of tomorrow.

And yet, as important (and as interesting) as it is, I have to add warfare and weaponry to the long list of important subjects not discussed (or at least, not discussed at any length) in this book. The use of technology in warfare is an excellent example of the thesis of this book, and so complex and far-reaching that it obviously deserves, perhaps *requires*, its own volume—several volumes, in fact. (A couple of the more interesting and accessible ones are Martin van Creveld's *Technology and War*, and, for an in-depth look at the historical foundations of weaponry, Alfred Bradford's *With Arrow, Sword, and Spear*.)

Of course, that could—and *should*—be said of any of the technologies I considered but ended up not including. Many of them deserve in-depth treatment of the sort that this volume does not attempt. And some of the gaps are obvious: I spoke of computers, but never examined Grace Hopper's pioneering role in programming either the Harvard Mark I or Mauchly and Eckert's UNIVAC. I wrote about Martin Cooper, the inventor of the cellular phone, and William Shockley, credited with the invention of the transistor, but certainly those brief mentions did not do justice to either man—or to the team of largely unsung men and women behind them who also played a large part in those inventions. There were dozens of subjects I would have loved to include but was unable to, mainly due to limitations of time or space: robotics, the potential uses of graphene, augmented reality and virtual reality, the telescope and microscope, the wheel and axle, the satellite, and so many others.

The list of overlooked technologies goes on and on, essentially forever, because technology itself is endless and invention knows no bounds. As I'm writing these words, new technologies are being born, and new uses are being found for existing technologies. It's obvious that there's no way to do justice to every technology in one book, especially a book that attempts nothing more than a cursory—but hopefully thought-provoking—overview of selected technologies, the development of which speak to the theme of the book.

This is not, after all, a survey of "all" technologies. Or the "most important" ones. Or even the newest, coolest, most interesting, or most notable ones. Instead, it's been an attempt to illustrate a theme by discussing, somewhat informally, some of the tools that are representative of that theme: the idea that most technologies begin life in some restricted form, that they are initially not available to the average person. That instead, such tools start off—and, for a period of time, remain—in the hands of some elite group: the wealthy, the powerful, the ruling class, the intellectually gifted, the well-connected. For one reason or another, access to much technology is at the outset circumscribed, reserved for some group outside of the mainstream. And often for good reason.

But eventually, that access trickles down. Always. Sometimes that happens rather quickly, and sometimes the trickle becomes a flood, but almost always there's some gap, some delay between the time a technology becomes available to some and the time it becomes available to all—or at least, to a very large number. (The thought can be a bit frightening; if true, it means that there will come a day when groups that are not states or state-sponsored—not that those aren't frightening enough—will get their hands on armed drones, for instance. Or on atomic weaponry. While we can try to restrict access to certain tools or technologies, it's difficult or impossible to do so indefinitely. Sooner or later, even dangerous technologies begin to trickle down. Technology is, after all, mainly just information, and information is an ethereal, insubstantial thing that can be contained only with difficulty, and then, not for very long.)

So, rather than attempting to provide an in-depth history of technology, the goal here was to create an illuminating and readable book that addressed the theme and that was also somewhat informal and conversational. I hope I succeeded, and I hope you found the book useful, enjoyable, and informative; to the degree that I failed, the fault is surely mine, and not that of any of the many people who contributed to the book.

Bibliography

Print and Internet

"15 Year Old Who 'SWATTED' Gamer Convicted of Domestic Terrorism; 25 Years to Life in Federal Prison." National Report RSS. August 30, 2014. Accessed March 20, 2015. http://nationalreport.net/15-year-old-swatted-domestic-terrorism/.

"100 Years of U.S. Consumer Spending: Data for the Nation, New York City, and Boston." US Bureau of Labor Statistics. August 3, 2006. Accessed April 21, 2015. http://www.bls.gov/opub/uscs/1918-19.pdf.

"3D Systems Awarded for Its Personalized 3D Printing Healthcare Solutions." Microfabricator.com. Microfabricator, 7 July 2014. Web. 21 September 2014. http://microfabricator.com/articles/view/id/53bad5463139448c128b4641/3d-systems-awarded-for-its-personalized-3d-printing-healthcare-solutions.

"3-Year-Old Maui Boy Receives 'Ironman'-like 3-D Printed Hand." YouTube. KHON2 News, 5 September 2014. Web. 20 September 2014. https://www.youtube.com/watch?v=CQKcEhTd-Ck&sns=em.

"2013 Ford Fiesta." Reviews, Specs and Prices. Accessed April 21, 2015. http://www.cars.com/ford/fiesta/2013/.

"2015 TRUSTe US Consumer Confidence Index." 2015 US Consumer Data Privacy Study: Consumer Privacy Edition from TRUSTe. Accessed January 30, 2015. http://www.truste.com/us-consumer-confidence-index-2015/.

Ackerman, Spencer. "Obama Must Finally End NSA Phone Record Collection, Says Privacy Board." *The Guardian*. January 29, 2015. Accessed January 30, 2015. http://www.theguardian.com/us-news/2015/jan/29/obama-end-nsa-phone-records-collection-privacy-board.

"The Age of the Automobile." UShistory.org. Accessed April 15, 2015. http://www.ushistory.org/us/46a.asp.

Alba, Davey. "This Device Diagnoses Hundreds of Diseases Using a Single Drop of Blood | WIRED." Wired.com. November 8, 2014. Accessed November 11, 2014. http://www.wired.com/2014/11/device-diagnoses-hundreds-diseases-using-single-drop-blood/.

Albright, Dann. "Avoiding Internet Surveillance: The Complete Guide." MakeUseOf. January 25, 2015. Accessed January 26, 2015. http://www.makeuseof.com/tag/avoiding-internet-surveillance-complete-guide/.

———. "What Can Government Security Agencies Tell from Your Phone's Metadata?" MakeUseOf. February 2, 2015. Accessed February 4, 2015. http://www.makeuseof.com/tag/can-government-security-agencies-tell-phones-metadata/.

Alonso-Zaldivar, Ricardo, and Jack Gillum. "New Privacy Concerns over Health Care Website." *Detroit News* Website. January 20, 2015. Accessed January 20, 2015. http://www.detroitnews.com/story/news/politics/2015/01/20/health-overhaul-privacy/22035873/.

Ambrose, Stephen E. *Undaunted Courage: Meriwether Lewis, Thomas Jefferson, and the Opening of the American West*. New York: Simon & Schuster, 1996.

Ampofo, Lawrence. "5 Ways the Internet of Things Will Change Social Media." Business 2 Community. 24 October 2014. Web. 25 October 2014. http://www.business2community.com/social-media/5-ways-internet-things-will-change-social-media-01047822.

Anderson, Janna Quitney, and Harrison Rainie. *Digital Life in 2025: Net Threats*. Washington, DC: Pew Research Center, 2014.

———. *Digital Life in 2025: The Internet of Things Will Thrive by 2025*. Washington, DC: Pew Research Center, 2014.

Anderssen, Erin. "Big Data Is Watching You. Has Online Spying Gone Too Far?" *Globe and Mail*. October 2, 2014. Accessed December 11, 2014. http://www.theglobeandmail.com/life/relationships/big-data-is-watching-you-has-online-spying-gone-too-far/article20894498/?page=all.

Andrews, Evan. "7 Things You May Not Know about the Gutenberg Bible." History.com. February 23, 2015. Accessed May 27, 2015. http://www.history.com/news/7-things-you-may-not-know-about-the-gutenberg-bible.

Anta, Rafael, Shireen El-Wahab, and Antonino Giuffrida. "Impact Pathways—MHealth." Mobile for Development Impact. Mobile for Development (M4D) Impact. Web. 18 August 2014. http://idbdocs.iadb.org/wsdocs/getdocument.aspx?docnum=1861959.

Armstrong, Karen. *The Bible: A Biography*. New York: Grove Press, 2008.

Atherton, Kelsey D. "The Week in Drones: Saving Lives, Spooking Nukes, and More." *Popular Science*. October 31, 2014. Accessed November 1, 2014. http://www.popsci.com/article/technology/week-drones-saving-lives-spooking-nukes-and-more.

Atiyeh, Clifford. "Screen-Plate Club: How License-Plate Scanning Compromises Your Privacy—Feature." *Car and Driver*. October 1, 2014. Accessed March 29, 2015. http://www.caranddriver.com/features/screen-plate-club-how-license-plate-scanning-compromises-your-privacy-feature.

"Automobiles." History.com. Accessed April 20, 2015. http://www.history.com/topics/automobiles.

Aznar, Miguel F. *Technology Challenged: Understanding Our Creations & Choosing Our Future*. Santa Cruz, CA: KnowledgeContext, 2005.

Badger, Emily. "This Is How Women Feel about Walking Alone at Night in Their Own Neighborhoods." *Washington Post*. 28 May 2014. Web. 12 October 2014. http://www.washingtonpost.com/blogs/wonkblog/wp/2014/05/28/this-is-how-women-feel-about-walking-alone-at-night-in-their-own-neighborhoods/.

Baehr, Leslie. "How Does GPS Work?" Seattle PI 23 July 2014, US ed., Business sec.: N/a. SeattlePI.com. Web. 22 August 2014. http://www.seattlepi.com/default/article/How-Does-GPS-Work-5641436.php.

Baker, Deirdre. "65th Anniversary of Historic Q-C Fire Is Wednesday." *Quad-City Times*. January 6, 2015. Accessed April 10, 2015. http://qctimes.com/news/local/th-anniversary-of-historic-q-c-fire-is-wednesday/article_ed244ee4-7c3d-5965-91ce-6c012621efec.html.

Bargmann, Joe. "What the Hummingbird Can Tell Us about Innovation." *Popular Mechanics*. October 15, 2014. Accessed April 18, 2015. http://www.popularmechanics.com/science/animals/a11385/what-the-hummingbird-can-tell-us-about-innovation-17310951/.

Barrat, James. "Why Stephen Hawking and Bill Gates Are Terrified of Artificial Intelligence." *Huffington Post / World Post*. April 9, 2015. Accessed April 10, 2015. http://www.huffingtonpost.com/james-barrat/hawking-gates-artificial-intelligence_b_7008706.html.

Barrett, Devlin. "Americans' Cellphones Targeted in Secret Spy Program." MarketWatch. November 13, 2014. Accessed March 22, 2015. http://www.marketwatch.com/story/americans-cellphones-targeted-in-secret-spy-program-2014-11-13.

Baugh, Dick, Vittorio Brizzi, and Tim Baker. "Ötzi's Bow." Ötzi's Bow. Primitiveways.com, 20 January 2006. Web. 9 August 2014. http://www.primitiveways.com/Otzi's_bow.html.

Beaumont-Thomas, Ben. "Ashley Judd to Press Charges against Twitter Users over Sexual Harassment." *The Guardian*. March 18, 2015. Accessed March 20, 2015. http://www.theguardian.com/film/2015/mar/18/ashley-judd-to-press-charges-twitter-sexual-harassment.

Behrman, Max. "Boeing's Phantom Ray Stealth UAV Makes First Flight." Defense Tech RSS. Gizmodo, 3 May 2011. Web. 14 September 2014. http://defensetech.org/2011/05/03/boeings-phantom-ray-stealth-uav-makes-first-flight/.

Bennett, Brian. "Spy Agencies Add Restrictions to Use of Email and Cellphone Data." *Los Angeles Times*. February 3, 2015. Accessed February 4, 2015. http://www.latimes.com/nation/la-na-spy-agencies-restrict-use-of-email-and-cell-phone-data-20150203-story.html.

"Benz Patent Motor Car, the First Automobile (1885–1886)." Daimler. Accessed April 21, 2015. http://www.daimler.com/dccom/0-5-1322446-1-1323352-1-0-0-1322455-0-0-135-0-0-0-0-0-0-0-0.html.

Berners-Lee, Tim, and Mark Fischetti. *Weaving the Web: The Original Design and Ultimate Destiny of the World Wide Web by Its Inventor*. San Francisco: HarperSanFrancisco, 1999.

Bernstein, Joseph. "Publishers Know You Didn't Finish 'The Goldfinch'—Here's What That Means for the Future of Books." BuzzFeed. January 21, 2015. Accessed February 9, 2015. http://www.buzzfeed.com/josephbernstein/publishers-know-you-didnt-finish-the-goldfinch-heres-what-th#.usAykd9bv.

"BlackBerry Timeline: A Look Back at the Tech Company's History." Global News and The Canadian Press. September 24, 2013. Accessed March 5, 2015. http://globalnews.ca/news/860689/blackberry-timeline-a-look-back-at-the-tech-companys-history/.

"Bliss Is in Contempt." *The Morning Record*, January 19, 1896, Meriden, CT.

Blue, Violet. "Tracker, Pixel, Spyware, Snitch: How to Shop in Secret | ZDNet." ZDNet. December 12, 2014. Accessed December 12, 2014. http://www.zdnet.com/article/tracker-pixel-spyware-snitch-how-to-shop-in-secret/.

Boaz, Noel Thomas, and Russell L. Ciochon. *Dragon Bone Hill: An Ice-Age Saga of Homo Erectus*. Oxford: Oxford University Press, 2004.

Boorstin, Daniel J. *The Creators*. New York: Random House, 1992.

Borenstein, Seth, and Jack Gillum. "Credit Card Data Not So Private." U-T San Diego. January 30, 2015. Accessed January 30, 2015. http://www.utsandiego.com/news/2015/jan/30/tp-credit-card-data-not-so-private/.

Borsa, Adrian Antal, Duncan Carr Agnew, and Daniel R. Cayan. "Ongoing Drought-Induced Uplift in the Western United States." Science Express 10.1126 (2014): N/a. *Science* Magazine. 2014. Web. 23 August 2014. http://www.sciencemag.org/content/early/2014/08/20/science.1260279.full.

Bort, Julie. "Here's How to Figure out Everything Google Knows about You." *Business Insider*. November 15, 2014. Accessed November 18, 2014. http://www.businessinsider.com/what-google-knows-about-you-2014-11.

"Boy Gets Prosthetic Hand Made by 3-D Printer." CBS Evening News, 28 October 2013. Web. 20 September 2014. https://www.youtube.com/watch?v=FGSo_I86_lQ.

boyd, danah. *It's Complicated: The Social Lives of Networked Teens*. New Haven, CT: Yale University Press, 2014.

Bradford, Alfred S., and Pamela M. Bradford. *With Arrow, Sword, and Spear: A History of Warfare in the Ancient World*. Westport, CT: Praeger, 2001.

"Breach of Promise." *Chicago Tribune*, January 12, 1883.

"Brief History of the Internet." Internet Society. Accessed April 5, 2015. http://www.internetsociety.org/internet/what-internet/history-internet/brief-history-internet.

Brustein, Joshua. "RadioShack's Bankruptcy Could Give Your Customer Data to the Highest Bidder." Bloomberg.com. March 24, 2015. Accessed March 25, 2015. http://www.bloomberg.com/news/articles/2015-03-24/radioshack-s-bankruptcy-could-give-your-customer-data-to-the-highest-bidder.

Brynjolfsson, Erik, and Andrew McAfee. *The Second Machine Age: Work, Progress, and Prosperity in a Time of Brilliant Technologies*. New York: W. W. Norton, 2014.

Bryson, Bill. *A Short History of Nearly Everything*. New York: Broadway, 2003.

———. *The Mother Tongue: English & How It Got That Way*. New York: Avon Books, 1991.

Burrus, Daniel. "3D Printed Shoes: A Step in the Right Direction | WIRED." Wired.com. Conde Nast Digital / Wired Magazine, 14 September 2010. Web. 20 September 2014. http://www.wired.com/2014/09/3d-printed-shoes/.

Burt, David, Aaron Kleiner, J. Paul Nichols, and Kevin Sullivan. "Navigating the Future of Cybersecurity Policy." Cyberspace 2025: Today's Decisions, Tomorrow's Terrain 1.1 (2014): 1-51. Microsoft Cyberspace2025. 2014. Web. 22 August 2014. http://www.microsoft.com/security/cybersecurity/cyberspace2025/.

Canavero, Sergio. "The 'Gemini' Spinal Cord Fusion Protocol: Reloaded." *Surgical Neurology International*. Wolters Kluwer 2015; 6:18

Carr, David. "Unease for What Microsoft's HoloLens Will Mean for Our Screen-Obsessed Lives." *New York Times*. January 25, 2015. Accessed January 26, 2015. http://www

.nytimes.com/2015/01/26/business/media/unease-for-what-microsofts-hololens-will-mean-for-our-screen-obsessed-lives.html.

Cate, Fred H. "Personal Privacy Is Eroding as Consent Policies of Google and Facebook Evoke 'Fantasy World.'" The Conversation. December 15, 2014. Accessed December 15, 2014. http://theconversation.com/personal-privacy-is-eroding-as-consent-policies-of-google-and-facebook-evoke-fantasy-world-33210.

"CGP Catalog—Basic Search: Full Catalog." Catalog of US Government Publications. Accessed May 23, 2015. http://catalog.gpo.gov/F?RN=858894529.

Chase, Robin. "Car-Sharing Offers Convenience, Saves Money and Helps the Environment." US Department of State, Bureau of International Information Programs. Accessed May 14, 2015. http://photos.state.gov/libraries/cambodia/30486/Publications/everyone_in_america_own_a_car.pdf.

Chen, Te-Ping, Fiona Law, and Newley Purnell. "Apps Speed Up, and Often Muddle, Hong Kong Protesters' Messages." *Wall Street Journal*. Dow Jones & Company, 9 October 2014. Web. 12 October 2014. http://online.wsj.com/articles/whatsapp-key-to-quickly-rallying-protesters-in-hong-kong-but-groups-struggle-to-stay-on-message-1412878808.

"Child's Play." Child's Play. Web. 12 October 2014. http://www.childsplaycharity.org/.

Christiansen, Trevor. "Why Hackers Love LinkedIn." Sungard AS Blog. 25 August 2014. Web. 10 October 2014. http://blog.sungardas.com/2014/08/is-your-linkedin-profile-putting-your-company-at-risk/#sthash.FTDBhcei.xGh6EpaD.dpbs.

City Record: Official Publication of the City of Boston. 21st ed. Vol. 10. Boston: City of Boston, 1918.

Clark, J. D., and J. W. K. Harris. "Fire and Its Roles in Early Hominid Lifeways." *African Archaeological Review*, 1985, 3–27.

Clark, Kenneth. *Civilisation: A Personal View*. 1st US ed. New York: Harper & Row, 1970, 1969.

Clark, Virginia P., Paul A. Eschholz, and Alfred F. Rosa, eds. *Language: Introductory Readings*. 4th ed. New York: St. Martin's Press, 1985.

Collins, Tom. *The Legendary Model T Ford: The Ultimate History of America's First Great Automobile*. Iola, WI: Krause Publications, 2007.

Conde Nast Digital. "The 15 Most Dangerous People in the World," Wired.com. http://www.wired.com/2012/12/most-dangerous-people/. Accessed September 23, 2014.

"Congolese Computer Inventor." TV2 Africa / Voice of America, 27 April 2012. Web. 18 September 2014. https://www.youtube.com/watch?v=37Xw0yad-D8.

"The Connected Car and Privacy: Navigating New Data Issues." The Future of Privacy Forum. November 13, 2014. Accessed November 16, 2014. http://www.futureofprivacy.org/wp-content/uploads/FPF_Data-Collection-and-the-Connected-Car_November2014.pdf.

Cook, William R., and Ronald B. Herzman. *The Medieval World View: An Introduction*. New York: Oxford UP, 1983.

Coren, Stanley. *How Dogs Think: What the World Looks Like to Them and Why They Act the Way They Do*. New York: Free Press, 2005.

Cuenta, Kristel. "4 Ways Social Media Is Utilized for Social Good." *Search Engine Journal.* 18 August 2014. Web. 12 October 2014. http://www.searchenginejournal.com/4-ways-social-media-utilized-social-good/111332/.

Curtis, Sophie. "Meet Amelia: The Computer that's after Your Job." *The Telegraph.* September 29, 2014. Accessed January 15, 2015. http://www.telegraph.co.uk/technology/news/11123336/Meet-Amelia-the-computer-thats-after-your-job.html.

daCosta, Francis. *Rethinking the Internet of Things: A Scalable Approach to Connecting Everything.* New York: ApressOpen, 2013.

Davies Boren, Zachary. "There Are Officially More Mobile Devices than People in the World." *The Independent.* Independent Digital News and Media, 2 October 2014. Web. 29 October 2014. http://www.independent.co.uk/life-style/gadgets-and-tech/news/there-are-officially-more-mobile-devices-than-people-in-the-world-9780518.html.

Davis, Kenneth C. *Don't Know Much about History: Everything You Need to Know about American History, but Never Learned.* New York: Crown, 1990.

Davis, Lauren. "10 Things People Once Complained Would Ruin the English Language." Io9. February 6, 2015. Accessed March 21, 2015. http://io9.com/10-things-people-once-complained-would-ruin-the-english-1684240298.

Della Cava, Marco. "Voices: In the Digital World, Privacy Is the Price of Admission." *USA Today.* 8 July 2014, Tech sec.: N/a.

———. "GoogleX Takes to the Skies with Secret Drone Project." *USA Today.* 28 August 2014, Technology sec.: N/a.

De Montjoye, Yves-Alexander, Laura Radaelli, Vivek Kumar Singh, and Alex Pentland. "Unique in the Shopping Mall: On the Reidentifiability of Credit Card Metadata." *Science.* January 30, 2015, 536–39.

Dendy, W. Mark. "A Web-Enabled Toothbrush Helps Keep Teeth White." Digital Journal. March 4, 2014. Accessed November 2, 2014. http://www.digitaljournal.com/technology/a-blue-tooth-toothbrush-helps-keep-teeth-white/article/374399.

Diamond, Jared M. *Guns, Germs, and Steel: The Fates of Human Societies.* New York: W. W. Norton, 1998.

Dilanian, Ken. "AP Exclusive: Before Snowden, a Debate inside NSA." The Big Story. November 19, 2014. Accessed November 20, 2014. http://bigstory.ap.org/article/acc54fc0c64c4c3eae29b8ac380cc065/ap-exclusive-snowden-debate-inside-nsa.

Dobra, Susan. "The Gift of Theuth: Plato on Writing (Again)." CSU Chico. Accessed November 9, 2014. http://www.csuchico.edu/phil/sdobra_mat/platopaper.html.

Doctorow, Cory. "Peak Indifference-to-surveillance." Boing Boing. November 12, 2014. Accessed November 20, 2014. http://boingboing.net/2014/11/12/peak-indifference-to-surveilla-2.html.

Doughton, Sandi. "Scientists Track Motions of Shifting Plates Using GPS Sensors." Scientists Track Motions of Shifting Plates Using GPS Sensors. Phys.org, 1 July 2011. Web. 23 August 2014. http://phys.org/news/2011-06-scientists-track-motions-shifting-plates.html.

"DRONES: Attack of the Drones." 4-Traders. November 22, 2014. Accessed November 23, 2014. http://www.4-traders.com/GOPRO-INC-16783944/news/GoPro—

DRONES-DRONES-DRONES-DRONES-Attack-of-the-drones-Little-hover ing-menaces-are-set-to-be-19438048/.

Dube, Ryan. "Top 5 Ways You Are Spied On Every Day And Don't Know It." MakeUse Of. December 19, 204. Accessed December 20, 2014. http://www.makeuseof.com/tag/top-5-ways-spied-every-day-dont-know/.

Ducklin, Paul. " 'Cheaper Car Insurance' Dongle Could Lead to a Privacy Wreck." Naked Security. January 20, 2015. Accessed January 25, 2015. https://nakedsecurity.sophos .com/2015/01/20/cheaper-car-insurance-dongle-could-lead-to-a-privacy-wreck/.

Duggan, Maeve. "5 Facts about Online Harassment." Pew Research Center RSS. October 30, 2014. Accessed April 12, 2015. http://www.pewresearch.org/fact-tank/2014/10/30 /5-facts-about-online-harassment/.

Dyson, George. *Turing's Cathedral: The Origins of the Digital Universe*. New York: Pantheon, 2012.

"Editorial: A Helping Hand by Air, Thanks to Civilian Drones." *Ventura County Star*. November 29, 2014. Accessed November 30, 2014. http://www.vcstar.com/opinion/editorials/editorial-a-helping-hand-by-air-thanks-to-civilian-drones_35075375.

Eisenstein, Paul. "Hot Trend in Automobiles: Not Owning One." CNBC. May 24, 2013. Accessed May 14, 2015. http://www.cnbc.com/id/100762511.

Elliott, Dan. "DesMoinesRegister.com | News from the Associated Press." DesMoinesReg ister.com | News from the Associated Press. December 18, 2014. Accessed December 18, 2014. http://hosted.ap.org/dynamic/stories/U/US_TORNADO_RESEARCH _DRONES_QA?SITE=IADES&SECTION=HOME&TEMPLATE=DEFAULT &CTIME=2014-12-18-00-27-04.

Ellul, Jacques, and John Wilkinson. *The Technological Society*. New York: Vintage, 1964.

Etherington, Darrell. "Google Acquires Titan Aerospace, The Drone Company Pursued By Facebook." TechCrunch. April 14, 2014. Accessed November 16, 2014. http://tech crunch.com/2014/04/14/google-acquires-titan-aerospace-the-drone-company-pur sued-by-facebook/.

"Etienne Lenoir | Biography." Encyclopedia Britannica Online. Accessed April 27, 2015. https://www.britannica.com/EBchecked/topic/336021/Etienne-Lenoir.

"FAA Aerospace Forecast: Fiscal Years 2010–2030." Federal Aviation Administration. January 1, 2010. Accessed November 16, 2014. http://www.faa.gov/data_research/aviation/aerospace_forecasts/2010-2030/media/2010 Forecast Doc.pdf.

"Famous Thanksgiving Weather." Earth Gauge—A National Environmental Education Foundation Program. Accessed April 4, 2015. http://www.earthgauge.net/wp-con tent/EG_Thanksgiving Weather.pdf.

Farr, Christina. "Brain-Training Game Maker Settles with U.S. FTC over Unsupported Claims." Reuters. January 20, 2015. Accessed March 5, 2015. http://www.reuters .com/article/2015/01/21/us-ftc-health-apps-idUSKBN0KU03B20150121.

Febvre, Lucien, and Henri Martin. *The Coming of the Book: The Impact of Printing 1450–1800*. London: NLB, 1976.

"Field Listing: Literacy." Central Intelligence Agency: World Factbook. Accessed May 17, 2015. https://www.cia.gov/library/publications/the-world-factbook/fields/print_2103 .html.

"Firelight Talk of the Kalahari Bushmen." September 22, 2014. Accessed January 6, 2015. http://unews.utah.edu/news_releases/firelight-talk-of-the-kalahari-bushmen/.

Ford, Henry, and Samuel Crowther. *My Life and Work.* New York: Arno Press, 1973.

"Ford Model T History and the Early Years of the Ford Motor Company." History of the Ford Model T, Henry Ford and Ford Motor Company Background, History and Facts. Accessed April 26, 2015. http://www.modelt.ca/background.html.

Fowler, Brenda. "The Iceman's Last Meal." PBS. Web. 8 August 2014. http://www.pbs.org/wgbh/nova/ancient/iceman-last-meal.html.

Franco, Michael. "Modern Magic: World's First 3D-Printed Castle—CNET." CNET. 28 August 2014. Web. 15 September 2014. http://www.cnet.com/news/worlds-first-3d-printed-castle-built-house-next/.

Friedman, Thomas L. *The World Is Flat: A Brief History of the Twenty-First Century.* New York: Farrar, Straus and Giroux, 2005.

Fung, Brian. "A Privacy Policy for Cars: What Automakers Know About You." *Daily Herald.* December 14, 2014. Accessed December 15, 2014. http://www.dailyherald.com/article/20141214/business/141219879/.

Gaeng, Paul A. *Introduction to the Principles of Language.* New York: Harper & Row, 1971.

Garcia, Zach. "What Flies When It Comes to Drone Laws Across the Globe." The Missouri Drone Journalism Program. April 19, 2013. Accessed December 21, 2014. http://www.missouridronejournalism.com/2013/04/what-flies-when-it-comes-to-drone-laws-across-the-globe/.

Gay, Peter. *The Enlightenment: An Interpretation.* 1st ed. New York: W. W. Norton, 1996.

Gellman, Barton, and Ashkan Soltano. "NSA Surveillance Program Reaches 'Into the Past' to Retrieve, Replay Phone Calls." *Washington Post,* March 22, 2014. Accessed March 22, 2015. http://www.washingtonpost.com/world/national-security/nsa-surveillance-program-reaches-into-the-past-to-retrieve-replay-phone-calls/2014/03/18/226d2646-ade9-11e3-a49e-76adc9210f19_story.html

Ghena, Branden, William Beyer, Jonathan Pevarnek, and J. Alex Halderman. "Green Lights Forever: Analyzing the Security of Traffic Infrastructure." Proceedings of the 8th USENIX Workshop on Offensive Technologies N/A. August (2014): 10.

Ghose, Tia. "Ötzi 'the Iceman' Had Heart Disease Genes." Fox News. FOX News Network, 31 July 2014. Web. 8 August 2014. http://www.foxnews.com/science/2014/07/31/otzi-iceman-had-heart-disease-genes/.

Gilbert, Jason. "Smartphone Usage Stats Suggest You Aren't Calling Your Mother (Or Anyone) Enough." *Huffington Post.* July 2, 2012. Accessed March 8, 2015. http://www.huffingtonpost.com/2012/07/02/smartphone-usage-stats_n_1643761.html.

Gilbert, Martin. *A History of the Twentieth Century.* New York: W. Morrow, 1997, 1999.

Gillespie, Alexandra. *The Production of Books in England 1350–1500.* Cambridge: Cambridge University Press, 2011.

Girard, Kim. "Internet of Things Presents CIOs with Both Technical and Ethical Questions." CIO. http://www.cio.com/article/2377877/privacy/internet-of-things-presents-cios-with-both-technical-and-ethical-questions.html (accessed September 24, 2014).

"Google Terms of Service—Privacy & Terms—Google." April 14, 2014. Accessed December 14, 2014. https://www.google.com/intl/en/policies/terms/.

Gorzelany, Jim. "Forget Phones or Fast Food, More Drivers Cause Their Own Distractions in Crashes." *Forbes* Magazine, 3 April 2013. Web. 14 September 2014. http://www.forbes.com/sites/jimgorzelany/2013/04/03/forget-phones-or-fast-food-more-drivers-cause-their-own-distractions-in-crashes/.

Goudsblom, Johan. "The Civilizing Process and the Domestication of Fire." *Journal of World History* 1992, Vol. 3, No. 1.

———. "The Domestication of Fire as a Civilizing Process." *Theory, Culture & Society*, 1992, 457–76.

"GPS Future and Evolutions." Navipedia. European Space Agency, 1 December 2011. Web. 1 September 2014. http://www.navipedia.net/index.php/GPS_Future_and_Evolutions.

Gray, Carroll. "Flying Machines—Samuel P. Langley." Accessed November 17, 2014. http://www.flyingmachines.org/lang.html.

Greenberg, Adam. "Mobile App Study Reveals Privacy Concerns." *SC Magazine*. 12 September 2014. Web. 15 September 2014. http://www.scmagazine.com/mobile-app-study-reveals-privacy-concerns/article/371312/.

———. "Report: More than 15,000 Vulnerabilities in Nearly 4,000 Applications Reported in 2014." *SC Magazine*. March 27, 2015. Accessed March 30, 2015. http://www.scmagazine.com/report-more-than-15000-vulnerabilities-in-nearly-4000-applications-reported-in-2014/article/405955/.

Greenberg, Andy. "How 3-D Printed Guns Evolved into Serious Weapons in Just One Year | WIRED." Wired.com. http://www.wired.com/2014/05/3d-printed-guns/. Accessed September 23, 2014.

Grisham, Lori. "Child Deaths in Hot Cars: 10 Key Facts." August 1, 2014. Accessed January 17, 2015. http://www.azcentral.com/story/news/nation-now/2014/07/02/child-deaths-hot-cars-facts/12055727/.

Gross, Doug. "Texas Company Makes Metal Gun with 3-D Printer." CNN. http://www.cnn.com/2013/11/08/tech/innovation/3d-printed-metal-gun/index.html (accessed September 23, 2014).

"Groundbreaking Hip and Stem Cell Surgery Completed Using 3D-Printed Implant." ScienceDaily, 16 May 2014. Web. 21 September 2014. http://www.sciencedaily.com/releases/2014/05/140516203334.htm.

"GSMA Intelligence—Analysis—Smartphone Forecasts and Assumptions 2007." Web. 29 October 2014. https://gsmaintelligence.com/analysis/2014/09/smartphone-forecasts-and-assumptions-20072020/443/.

"The Gutenberg Bible." The British Library. Accessed May 25, 2015. http://www.bl.uk/treasures/gutenberg/homepage.html.

"Gutenberg's Print Shop." Harry Ransom Center: The University of Texas at Austin. Accessed May 15, 2015. http://www.hrc.utexas.edu/educator/modules/gutenberg/invention/printshop/.

Hadnagy, Christopher. *Social Engineering: The Art of Human Hacking*. Indianapolis: Wiley, 2011.

Hadnagy, Christopher, and Paul Ekman. *Unmasking the Social Engineer: The Human Element of Security*. Hoboken: Wiley, 2014.

Hahn, Jason. "Pew: Internet Privacy Is a Fantasy, Will Merely Be a 'Fetish' by 2025, According to Experts." Digital Trends. December 20, 2014. Accessed December 21,

2014. http://www.digitaltrends.com/web/pew-internet-privacy-fantasy-will-merely-fetish-2025-according-experts/.

———. "Refurbished Smartphones Are Big Business, Global Market to Hit $14 Billion by 2017." Digital Trends. February 21, 2015. Accessed February 23, 2015. http://www.digitaltrends.com/mobile/refurbished-smartphones-are-big-business-global-market-to-hit-14-billion-by-2017/.

"Hailing the History of New York's Yellow Cabs." NPR. January 8, 2007. Accessed April 14, 2015. http://www.npr.org/templates/story/story.php?storyId=11804573.

Hales, Peter. "Levittown: Documents of an Ideal American Suburb." University of Illinois at Chicago. Accessed April 15, 2015. http://tigger.uic.edu/~pbhales/Levittown.html.

Hall, Stephen. "Iceman Autopsy." Pictures, More from *National Geographic* Magazine. National Geographic, 1 November 2011. Web. 8 August 2014. http://ngm.nationalgeographic.com/2011/11/iceman-autopsy/hall-text.

Halterman, Todd. "Ten 3D Printed Houses in a Day." 3D Printer World. 17 April 2014. Web. 28 September 2014. http://www.3dprinterworld.com/article/ten-3d-printed-houses-day.

Hanebutt-Benz, Eva-Maria. "Gutenberg Und Seine Zeit: Gutenberg Und Mainz." Gutenberg Und Seine Zeit: Gutenberg Und Mainz. Accessed May 17, 2015. http://www.gutenberg.de/english/zeitgum.htm.

Hare, Brian, and Vanessa Woods. *The Genius of Dogs: How Dogs Are Smarter Than You Think.* New York: Penguin/Plume Books, 2013.

Harris, Marvin. *Culture, People, Nature: An Introduction to General Anthropology*. 6th ed. New York: HarperCollins College, 1993.

Harris, Richard. "Why Typhoon Haiyan Caused So Much Damage." NPR. 1 November 2013. Web. 12 October 2014. http://www.npr.org/2013/11/11/244572227/why-typhoon-haiyan-caused-so-much-damage.

Harwell, Drew. "Whirlpool's Internet of Things Problem," *Washington Post*. 28 October 2014. Web. 29 October 2014. http://www.washingtonpost.com/blogs/the-switch/wp/2014/10/28/whirlpools-internet-of-things-problem-no-one-really-wants-a-smart-washing-machine/.

Hawley, E. Haven. "Economic Growth, Employment, and Access: A Case Study of Technological Change in the Printing Industry." Paper submitted for "Seminar on Industrial Modernization," Philip Shapira / Jan Youtie. Georgia Institute of Technology.

Headrick, Daniel R. *Technology: A World History*. Oxford: Oxford UP, 2009.

Hendel, John. "Celebrating Linotype, 125 Years Since Its Debut." *The Atlantic*. May 20, 2011. Accessed May 23, 2015. http://www.theatlantic.com/technology/archive/2011/05/celebrating-linotype-125-years-since-its-debut/238968/.

Heneghan, Tom. "Secrets behind the Forbidden Books." *America: The National Catholic Review*. February 7, 2005. Accessed May 27, 2015. http://americamagazine.org/issue/517/article/secrets-behindthe-forbidden-books.

Herbermann, Charles George. *The Catholic Encyclopedia: An International Work of Reference on the Constitution, Doctrine, Discipline, and History of the Catholic Church*. New York: Robert Appleton, 1907.

Herlihy, David. *Medieval Households*. Cambridge: Harvard University Press, 1985.

Hern, Alex. "Gamergate Hits New Low with Attempts to Send Swat Teams to Critics." *The Guardian*. January 13, 2015. Accessed February 1, 2015. http://www.theguardian.com/technology/2015/jan/13/gamergate-hits-new-low-with-attempts-to-send-swat-teams-to-critics.

———. "Samsung Rejects Concern over 'Orwellian' Privacy Policy." *The Guardian*. February 9, 2015. Accessed February 9, 2015. http://www.theguardian.com/technology/2015/feb/09/samsung-rejects-concern-over-orwellian-privacy-policy.

Hicks, Tony. "Emma Watson Says She Was Threatened after U.N. Speech." *San Jose Mercury News*. March 19, 2015. Accessed March 20, 2015. http://www.mercurynews.com/celebrities/ci_27676320/emma-watson-says-she-was-threatened-after-u.

"Highway History." Federal Highway Administration: The Interstate 50th Anniversary Web Site. Accessed April 29, 2015. http://www.fhwa.dot.gov/interstate/homepage.cfm.

Hill, Kashmir. "Hertz Puts Cameras in Its Rental Cars, Says It Has No Plans to Use Them." *Fusion*. March 16, 2015. Accessed March 16, 2015. http://fusion.net/story/61741/hertz-cameras-in-rental-cars/.

———. "Ten Fun Facts about Drones." *Forbes*. February 9, 2012. Accessed November 16, 2014. http://www.forbes.com/sites/kashmirhill/2012/02/09/10-fun-facts-about-drones/.

Hines, Pierre. "Learn to Stop Worrying and Love the Drones." *The Daily Beast*. September 17, 2013. Accessed November 16, 2014. http://www.thedailybeast.com/articles/2013/09/17/learn-to-stop-worrying-and-love-the-drones.html.

"The History and Evolution of Cell Phones." The Art Institutes Blog. Accessed March 5, 2015. http://new.artinstitutes.edu/Blog/the-history-and-evolution-of-cell-phones.

Holley, Peter. "Apple Co-Founder on Artificial Intelligence: 'The Future Is Scary and Very Bad for People.'" *Washington Post*. March 24, 2015. Accessed April 10, 2015. http://www.washingtonpost.com/blogs/the-switch/wp/2015/03/24/apple-co-founder-on-artificial-intelligence-the-future-is-scary-and-very-bad-for-people/.

"Home Computer Access and Internet Use." Child Trends. Web. 18 October 2014. http://www.childtrends.org/?indicators=home-computer-access.

"Homo Erectus." History.com. Accessed January 13, 2015. http://www.history.com/videos/homo-erectus#homo-erectus.

Hopewell, Luke. "3D Printer Can Build You a House in 20 Hours: Welcome to the Future." Gizmodo Australia. Gizmodo, 9 August 2012. Web. 14 September 2014. http://www.gizmodo.com.au/2012/08/3d-printer-can-build-you-a-house-in-20-hours-welcome-to-the-future/.

———. "How Much Money Do Tech Giants like Apple, Google, Microsoft and Others Make Every Minute?" Gizmodo Australia. Gizmodo / Allure Media, 19 March 2014. Web. 5 October 2014. http://www.gizmodo.com.au/2014/03/how-much-money-do-tech-giants-like-apple-google-microsoft-and-others-make-every-minute/.

Howard, Alexander B. "Whose Privacy Will Uber Violate Next? Why Its Latest Bad Behavior Matters | WIRED." Wired.com. November 16, 2014. Accessed November 19, 2014. http://www.wired.com/2014/11/can-we-trust-you-uber/.

"How Much Do Americans Earn? What Is the Average US Income and Other Income Figures. Fiscal Cliff Talks Only Useful in Context of Incomes." My Budget 360 RSS.

Accessed April 21, 2015. http://www.mybudget360.com/how-much-do-americans-earn-what-is-the-average-us-income/.
"How Was the Bible Distributed before the Printing Press Was Invented in 1455?" Biblica.com. December 27, 2013. Accessed May 17, 2015. http://www.biblica.com/en-us/bible/bible-faqs/how-was-the-bible-distributed-before-the-printing-press-was-invented-in-1455/.
Hughes, Cedric. "The Tragedy of Henry Bliss—The First of Many." *Road Rules*. October 11, 2011. Accessed April 14, 2015. http://www.roadrules.ca/content/tragedy-henry-bliss-first-many.
Hutchinson, Lee. "The Big Deal about 'Big Data'—Your Guide to What the Heck It Actually Means." ARS Technica. February 24, 2015. Accessed February 25, 2015. http://arstechnica.com/information-technology/2015/02/the-big-deal-about-big-data-your-guide-to-what-the-heck-it-actually-means/.
Hysing, Mari, Ståle Pallesen, Kjell Morten Stormark, Reidar Jakobsen, Astri J. Lundervold, and Børge Sivertsen. "Sleep and Use of Electronic Devices in Adolescence: Results from a Large Population-Based Study." BMJ Open. December 2, 2014. Accessed February 3, 2015. http://bmjopen.bmj.com/content/5/1/e006748.
Infante, Andre. "5 Disruptive Technology Breakthroughs that Will Shock the World." MakeUseOf. November 17, 2014. Accessed November 18, 2014. http://www.makeuseof.com/tag/5-disruptive-technology-breakthroughs-will-shock-world/.
"Infographic: The Future of EBooks." IDG Connect. 13 May 2014. Web. 17 August 2014. http://www.idgconnect.com/abstract/8177/infographic-the-future-ebooks.
"Internet of Things." *FTC Staff Report*, January, 2015. US Federal Trade Commission, Washington, DC.
"Internet of Things Research Study." *SC Magazine*. HP. Web. 15 August 2014. http://media.scmagazine.com/documents/88/hp_-_internet_of_things_21971.pdf.
"Interview: Ford's Anti-Semitism." PBS: *American Experience*. Accessed April 26, 2015. http://www.pbs.org/wgbh/americanexperience/features/interview/henryford-antisemitism/.
"Iraq's Ancient Past: Rediscovering Ur's Royal Cemetery." Penn Museum. Accessed February 8, 2015. http://www.penn.museum/sites/iraq/
Jackendoff, Ray. "How Did Language Begin?" Linguistic Society of America. August 1, 2006. Accessed February 15, 2015. http://www.linguisticsociety.org/files/LanguageBegin.pdf.
Jacobs, Rose. "Rise of Robot Factories Leading 'Fourth Industrial Revolution.'" Newsweek.com. March 5, 2015. Accessed March 6, 2015. http://www.newsweek.com/2015/03/13/rise-robot-factories-leading-fourth-industrial-revolution-311497.html.
Jarrett, Christian. "The Internet Probably Isn't Ruining Your Teenager's Brain | Science Blogs | WIRED." Wired.com. http://www.wired.com/2014/09/is-the-internet-scrambling-our-teenagers-brains-we-dont-know-but-probably-not/. Accessed September 4, 2014.
Jeffrey, Colin. "Telescopic Contact Lenses Zoom In and Out with the Wink of an Eye." Gizmag. February 16, 2015. Accessed February 16, 2015. http://www.gizmag.com/telescopic-contact-lenses-macular-degeneration-amd/36094/.

"Johannes Gutenberg." Biography.com. Accessed May 10, 2015. http://www.biography.com/people/johannes-gutenberg-9323828.

Johnson, Alan. "Drones the New Tool for Inspection and Audits in Heavy Industry." *Manufacturers Monthly*. November 21, 2014. Accessed November 21, 2014. http://www.manmonthly.com.au/news/drones-the-new-tool-for-inspection-and-audits-in-h.

Johnson, Paul. *The Birth of the Modern: World Society, 1815–1830*. New York: HarperCollins, 1991.

———. *Modern Times: The World from the Twenties to the Nineties*. Revised edition, 1st HarperPerennial ed. New York: HarperPerennial, 1992.

Johnson, Steven. *How We Got to Now: Six Innovations that Made the Modern World*. New York: Riverhead Books / Penguin Books, 2014.

Judd, Ashley. " 'Kiss My Ass': Ashley Judd Stands Up to Threats, Fights for Women Online." *Mic*. March 19, 2015. Accessed March 27, 2015. http://mic.com/articles/113226/forget-your-team-your-online-violence-toward-girls-and-women-is-what-can-kiss-my-ass.

Kacirk, Jeffrey. *Forgotten English*. New York: William Morrow, 1997.

Kaiser, Tiffany. "Court Rules that FAA Cannot Ban Commercial Drones, Dismisses $10,000 Fine for Drone User." DailyTech. March 7, 2014. Accessed November 9, 2014. http://www.dailytech.com/Court Rules that FAA Cannot Ban Commercial Drones Dismisses 10000 Fine for Drone User/article34468.htm.

Kaplan, Matt. "Million-Year-Old Ash Hints at Origins of Cooking." Nature.com. April 2, 2012. Accessed December 29, 2014. http://www.nature.com/news/million-year-old-ash-hints-at-origins-of-cooking-1.10372.

Karp, Aaron. "FAA: US Commercial Aircraft Fleet Shrank in 2011." ATWOnline. March 12, 2012. Accessed November 16, 2014. http://atwonline.com/aircraft-amp-engines/faa-us-commercial-aircraft-fleet-shrank-2011.

Katz, William A., and Svend Dahl. *Dahl's History of the Book*. 3rd English ed. Metuchen, NJ: Scarecrow Press, 1995.

Kaykas-Wolff, Jascha. "Why Privacy Is Like the Frog in the Pot of Boiling Water." Gigaom. November 29, 2014. Accessed November 30, 2014. https://gigaom.com/2014/11/29/why-privacy-is-like-the-frog-in-the-pot-of-boiling-water/.

Keenan, Thomas P. *Technocreep: The Surrender of Privacy and the Capitalization of Intimacy*. S.l.: Greystone, 2014.

Kelly, Heather. "Survey: Will We Give up Privacy without a Fight?" CNN. December 18, 2014. Accessed December 19, 2014. http://www.cnn.com/2014/12/18/tech/innovation/pew-future-of-privacy/.

Kintgen, Eugene R. *Perspectives on Literacy*. Carbondale, IL: Southern Illinois University Press, 1988.

Kiss, Jemima. " 'What Does ISP Mean?' How Government Officials Are Flunking Security Challenges." *The Guardian*. March 8, 2014. Accessed April 8, 2015. http://www.theguardian.com/technology/2014/mar/08/sxsw-government-cybersecurity-hacking-privacy-nsa-peter-singer.

Kistler, Robert E. "Fourth Presidential Policy Forum: Weather and National Security." Lecture, 84th AMS Annual Meeting from American Meteorological Society, Seattle, January 15, 2004.

Knight, Matt. "U.S. Planes Spy on American Cell Phones, Report Says." WTKRcom. Accessed November 14, 2014. http://wtkr.com/2014/11/14/u-s-planes-spy-on-american-cell-phones-report-says/.

Kohn, Alfie. *No Contest: The Case against Competition*. Boston: Houghton Mifflin, 1986.

Komando, Kim. "Q&A: Stop Advertisers from Tracking Your Phone." *USA Today*. March 20, 2015. Accessed March 22, 2015. http://www.usatoday.com/story/tech/columnist/komando/2015/03/20/qanda-advertiser-tracking/25024085/.

———. "Your Car's Hidden 'Black Box' and How to Keep It Private." *USA Today*. December 26, 2014. Accessed January 3, 2015. http://www.usatoday.com/story/tech/columnist/komando/2014/12/26/keep-your-car-black-box-private/20609035/.

Komorowski, Matt. "Matt Komorowski—A History of Storage Cost." Matt Komorowski—A History of Storage Cost. September 8, 2009. Accessed November 2, 2014. http://www.mkomo.com/cost-per-gigabyte.

Kong, Longbo, Zhi Liu, and Yan Huang. "SPOT: Locating Social Media Users Based on Social Network Context." Proceedings of the VLDB Endowment 7, No. 13 (2014): 1681–684.

Kravets, David. "U.N. Report Declares Internet Access a Human Right | WIRED." Wired.com. June 3, 2011. Accessed April 7, 2015. http://www.wired.com/2011/06/internet-a-human-right/.

Lang, C. A. "Software: Practice and Experience." Computer Typesetting. October 27, 2006. Accessed May 23, 2015. http://onlinelibrary.wiley.com/doi/10.1002/spe.4380060302/pdf.

Lascelles, Christopher. *A Short History of the World*. Revised edition. Dersingham, UK: Crux Publishing, 2014.

"Laser-Generated Surface Structures Create Extremely Water-Repellent, Self-Cleaning Metals." ScienceDaily. January 20, 2015. Accessed January 21, 2015. http://www.sciencedaily.com/releases/2015/01/150120111240.htm.

Leakey, Richard E., and Roger Lewin. *Origins: What New Discoveries Reveal about the Emergence of Our Species and Its Possible Future*. London: Macdonald and Janea's, 1977.

Lebergott, Stanley. "Wages and Working Conditions." *The Concise Encyclopedia of Economics*. Accessed April 20, 2015. http://www.econlib.org/library/Enc1/WagesandWorkingConditions.html#table 1.

LeClair, Dave. "How Is the Internet Changing the Traditional Workplace?" MakeUseOf. March 30, 2015. Accessed March 30, 2015. http://www.makeuseof.com/tag/how-is-the-internet-changing-the-traditional-workplace/.

Lerner, Robert E., Standish Meacham, and Edward McNall Burns. *Western Civilizations: Their History and Their Culture*, Vol. 1, 12th ed. New York: Norton, 1993.

Levin, Alan. "Ruling Bolsters FAA's Power to Regulate Civilian Drones." *Seattle Times*. November 18, 2014. Accessed November 19, 2014. http://seattletimes.com/html/businesstechnology/2025051382_faadronesxml.html.

Levine, Barry. "Look Out, Google—Intel Buys Chunk of Smart Glasses Maker Vuzix." VentureBeat. January 3, 2015. Accessed January 3, 2015. http://venturebeat.com/2015/01/03/look-out-google-intel-buys-chunk-of-smart-glasses-maker-vuzix/.

———. "Riding High, Adobe's Marketing Cloud Now Knows When You're in a Store." VentureBeat. November 17, 2014. Accessed November 18, 2014. http://venturebeat

.com/2014/11/17/riding-high-adobes-marketing-cloud-now-knows-when-youre-in-a-store/.

Leyden, John. "Stupid Humans and Their Expensive Data Breaches." *The Register*. December 5, 2014. Accessed December 8, 2014. http://www.theregister.co.uk/2014/12/05/stupid_humans_and_their_data_breaches/.

Linder, Robert Dean. *The Reformation Era*. Westport, CT: Greenwood Press, 2008.

"Lithography." PrintWiki. Accessed May 23, 2015. http://printwiki.org/Lithography.

Livingston, James D. "The Death of Evalina Bliss." *Arsenic and Clam Chowder: Murder in Gilded Age New York*. Albany, NY: Excelsior Editions, State University of New York Press, 2010.

Lohr, Steve. "Using Patient Data to Democratize Medical Discovery." Bits: Using Patient Data to Democratize Medical Discovery Comments. April 2, 2015. Accessed April 3, 2015. http://bits.blogs.nytimes.com/2015/04/02/using-patient-data-to-democratize-medical-discovery/.

Love, Cynthia D., Sean T. Lawson, and Avery E. Holton. "News from Above: First Amendment Implications of the Federal Aviation Administration Ban on Commercial Drones." Academia.edu. George Mason University. Web. 20 September 2014. https://www.academia.edu/8385487/News_from_Above_First_Amendment_Implications_of_the_Federal_Aviation_Administration_Ban_on_Commercial_Drones.

Luca, Michael. "Reviews, Reputation, and Revenue: The Case of Yelp.com." Harvard Business School. September 16, 2011. Accessed March 11, 2015. http://www.hbs.edu/faculty/Publication Files/12-016_0464f20e-35b2-492e-a328-fb14a325f718.pdf.

Lule, Jack. *Understanding Media and Culture: An Introduction to Mass Communication*. Irvington, NY: Flat World Knowledge, 2013.

Lunden, Ingrid. "1.2B Smartphones Sold in 2014, Led by Larger Screens and Latin America." TechCrunch. February 16, 2015. Accessed February 17, 2015. http://techcrunch.com/2015/02/16/1-2b-smartphones-sold-in-2014-led-by-larger-screens-and-latin-america/.

Luther, Martin, and Philipp Melanchthon. *Hymns of the Reformation*. London: Charles Gilpin, 1845.

Lynch, Tyler Wells. "Promise of 3D Printing May Be Overhyped." *USA Today*. 12 September 2013. Web. 14 September 2014. http://www.usatoday.com/story/tech/2013/09/11/reviewed-3d-printing-hype/2794041/.

Mackey, Robert. "Internet Star @ Least 473 Years Old." Internet Star @ Least 473 Years Old. The Lede / *New York Times* Blog, 4 May 2009. Web. 7 October 2014. http://thelede.blogs.nytimes.com/2009/05/04/internet-star-least-473-years-old/.

Madden, Mary, Amanda Lenhart, Sandra Cortesi, Urs Gasser, Maeve Duggan, Aaron Smith, and Meredith Beaton. "Teens, Social Media, and Privacy." Pew Research Centers Internet American Life Project RSS. Pew Research Center, 21 May 2013. Web. 15 October 2014. http://www.pewinternet.org/2013/05/21/teens-social-media-and-privacy/.

———. "Public Perceptions of Privacy and Security in the Post-Snowden Era." Pew Research Centers Internet American Life Project RSS. November 12, 2014. Accessed November 26, 2014. http://www.pewinternet.org/2014/11/12/public-privacy-perceptions/.

"Made in Space: 3D Printer Heading into Orbit." CBS News. CBS Interactive, 19 September 2014. Web. 21 September 2014. http://www.cbsnews.com/news/made-in-space-3d-printer-heading-into-orbit/.

Man, John. *Gutenberg: How One Man Remade the World with Words*. New York: John Wiley & Sons, 2002.

Manchester, William. *A World Lit Only by Fire: The Medieval Mind and the Renaissance: Portrait of an Age*. Boston: Little, Brown, 1992.

Marguccio, Nick. "1950 Thanksgiving Day Storm." November 21, 2012. Accessed April 1, 2015. http://www.weatherworksinc.com/thanksgiving-1950-storm.

Marshall, Michael. "Timeline: Weapons Technology." NewScientist. Reed Business Information, 7 July 2009. Web. 22 September 2014. http://www.newscientist.com/article/dn17423-timeline-weapons-technology.html#.VCCRsmeYaos.

Martin, Josh, David Myhre, and Nisha Singh. *Savings as a Cornerstone: Laying the Foundation for Financial Inclusion*. Arlington, VA: SEEP Network, 2014.

"Martin Luther." PBS. Accessed May 17, 2015. http://www.pbs.org/empires/martinluther/about_driv.html.

Maskey, Morag, Jessica Lowry, Jacqui Rodgers, Helen McConachie, and Jeremy R. Parr. "Reducing Specific Phobia/Fear in Young People with Autism Spectrum Disorders (ASDs) through a Virtual Reality Environment Intervention." *PLoS ONE* 9.PLoS ONE 9(7) (2014): N/A.

Mattessich, Richard. "Prehistoric Accounting and the Problem of Representation: On Recent Archaeological Evidence of the Middle-East from 8000 B.C. to 3000 B.C." *The Accounting Historians Journal*, October 1, 1987.

Matyszczyk, Chris. "Teens Love iPhone More, Use Facebook a Lot Less, Says Survey—CNET." CNET: Tech Culture. CNET, 7 October 2014. Web. Accessed 10 October 2014. http://www.cnet.com/news/teens-love-iphone-more-than-ever-but-iwatch-doesnt-excite/.

McCullough, David G. *The Wright Brothers*. New York: Simon & Schuster, 2015.

McQuaid, Hugh. "Committee Approves Drone Recommendations | CT News Junkie." CT News Junkie. December 12, 2014. Accessed December 14, 2014. http://www.ctnewsjunkie.com/archives/entry/committee_approves_drone_recommendations.

McWhorter, John. "Txtng is killing language. JK!!!" Lecture, 2014 TED Talk. February 1, 2013. http://on.ted.com/McWhorter

Meacham, Jon. *Thomas Jefferson: The Art of Power*. New York: Random House, 2012.

Mearian, Lucas. "3D Printing Techniques Will Be Used to Construct Buildings, Here and in Outer Space." *Computerworld*. IDG, 18 September 2013. Web. 28 September 2014. http://www.computerworld.com/article/2485200/emerging-technology/3d-printing-techniques-will-be-used-to-construct-buildings—here-and-in-outer-sp.html.

Merrill, Brad. "How Eel Drones Will Turn the Tide on Underwater Warfare." MakeUseOf. January 2, 2015. Accessed January 3, 2015. http://www.makeuseof.com/tag/eel-drones-will-turn-tide-underwater-warfare/.

Miller, Claire. "Americans Say They Want Privacy, but Act as If They Don't." *New York Times*. November 12, 2014. Accessed November 13, 2014. http://www.nytimes.com/2014/11/13/upshot/americans-say-they-want-privacy-but-act-as-if-they-dont.html.

Molitch-Hou, Michael. "University of Southern California & the Realization of 3D Printed Houses—3D Printing Industry." 3D Printing Industry. 30 September 2013. Web. Accessed 28 September 2014. http://3dprintingindustry.com/2013/09/30/university-south-california-realization-3d-printed-houses/.

Morgan, Jacob. "Privacy Is Completely and Utterly Dead, and We Killed It." *Forbes* Magazine, 9 August 2014. Web. 8 October 2014. http://www.forbes.com/sites/jacobmorgan/2014/08/19/privacy-is-completely-and-utterly-dead-and-we-killed-it/.

Morgan, James. "Forest Change Mapped by Google Earth." BBC News. 14 November 2013. Web. 31 August 2014. http://www.bbc.com/news/science-environment-24934790.

Morin, Monte. "When—and Where—Did Dogs First Become Our Pets?" *Los Angeles Times*. November 14, 2013. Accessed January 11, 2015. http://articles.latimes.com/2013/nov/14/science/la-sci-sn-dogs-domesticated-in-europe-20131114.

Mosbergen, Dominique. "WREX, 3D-Printed Exoskeleton, Helps Girl with Rare Congenital Disorder Move Her Arms (Video)." *Huffington Post*. TheHuffingtonPost.com, 3 August 2012. Web. 21 September 2014. http://www.huffingtonpost.com/2012/08/03/wrex-3d-printed-exoskeleton-girl-move-arms_n_1739419.html.

Moss, Richard. "Inkjet Printers Could Produce Paper Sensors that Identify Dangerous Food and Water Contaminants." Gizmag—Science. April 9, 2015. Accessed April 21, 2015. http://www.gizmag.com/inkjet-printed-paper-sensor/36915/.

The Motor. Vol. 4. London: Temple Press Limited, 1903.

Mott, Maryann. "Animal-Human Hybrids Spark Controversy." *National Geographic*. National Geographic Society, 25 January 2005. Web. 21 September 2014. http://news.nationalgeographic.com/news/2005/01/0125_050125_chimeras.html.

Moynihan, Tim. "Google Takes on the Challenge of Making Robot Surgery Safer | WIRED." Wired.com. March 30, 2015. Accessed March 30, 2015. http://www.wired.com/2015/03/google-robot-surgery/.

Mueffelmann, Kurt. "Uber's Privacy Woes Should Serve as a Cautionary Tale for All Companies | WIRED." Wired.com. January 23, 2015. Accessed January 24, 2015. http://www.wired.com/2015/01/uber-privacy-woes-cautionary-tale/.

"Multimedia: Charts." The Hamilton Project. Brookings Institution. Web. 18 October 2014. http://www.hamiltonproject.org/multimedia/charts/cost_of_computing_power_equal_to_an_ipad2/.

Murphy, Kate. "Things to Consider Before Buying that Drone." *New York Times*. December 6, 2014. Accessed December 7, 2014. http://www.nytimes.com/2014/12/07/sunday-review/things-to-consider-before-buying-that-drone.html.

Myers, L. M. *The Roots of Modern English*. Boston: Little, Brown, 1966.

Nair, Kusum. "Origins of Agriculture." Encyclopedia Britannica Online. March 16, 2014. Accessed February 15, 2015. http://www.britannica.com/EBchecked/topic/9647/origins-of-agriculture.

"Nanoribbon Film Keeps Glass Ice-free." ScienceDaily, 16 September 2014. Web. 17 September 2014. http://www.sciencedaily.com/releases/2014/09/140916155245.htm.

Nazaryan, Alexander. *Newsweek*. March 22, 2015. Accessed March 23, 2015. http://www.newsweek.com/youre-100-percent-wrong-about-privacy-315649.

Nest Labs, Inc. "Nest Protect Carbon Monoxide Field Study: Results from November 2013 to May 2014." Nest.com. https://nest.com/downloads/press/documents/co-white-paper.pdf. Accessed August 30, 2014.

Newman, Judith. "To Siri, With Love." *New York Times*. 17 October 2014. Web. 18 October 2014. http://www.nytimes.com/2014/10/19/fashion/how-apples-siri-became-one-autistic-boys-bff.html.

Newton, Jim. "Drones and the LAPD." *Los Angeles Times*. November 16, 2014. Accessed November 18, 2014. http://www.latimes.com/opinion/op-ed/la-oe-newton-column-lapd-drones-20141117-column.html.

Nissenbaum, Helen. *Privacy in Context: Technology, Policy, and the Integrity of Social Life*. Stanford, CA: Stanford Law Books, 2010.

"NTS-2 Satellite | Time and Navigation." NTS-2 Satellite | Time and Navigation. Smithsonian National Air & Space Museum. Web. 31 August 2014. http://timeandnavigation.si.edu/multimedia-asset/nts-2-satellite.

Ofek, Haim. *Second Nature: Economic Origins of Human Evolution*. Cambridge, UK: Cambridge University Press, 2001.

Overdorf, Jason. "Robots, Not Immigrants, Could Take Half of German Jobs." GlobalPost. January 12, 2015. Accessed January 15, 2015. http://www.globalpost.com/dispatch/news/regions/europe/germany/150112/robots-not-immigrants-could-take-half-german-jobs.

Paletz, David L., and Diana Marie Owen. *American Government and Politics in the Information Age*. Washington, DC: Flat World Knowledge, 2013.

Panati, Charles. *Extraordinary Origins of Everyday Things*. New York: Perennial Library, 1987.

Panchev, Stoĭcho. *Dynamic Meteorology*. Dordrecht: D. Reidel Pub., 1985.

"Patent Images." Patent Images—US Patent and Trademark Office. Accessed November 17, 2014. http://pdfpiw.uspto.gov/.piw?docid=00821393&SectionNum=1&IDKey=696CE58AE71C&HomeUrl=http://patft.uspto.gov/netahtml/PTO/patimg.htm.

Payack, Paul J. J. *A Million Words and Counting: How Global English Is Rewriting the World*. New York: Citadel, 2008.

Payton, Theresa, and Ted Claypoole. *Privacy in the Age of Big Data: Recognizing Threats, Defending Your Rights, and Protecting Your Family*. New York: Rowman & Littlefield, 2014.

Perera, David. "Smart Grid Powers up Privacy Worries." POLITICO. January 1, 2015. Accessed January 2, 2015. http://www.politico.com/story/2015/01/energy-electricity-data-use-113901.html.

Perry, Caroline. "Reengineering Privacy, Post-Snowden." Reengineering Privacy, Post-Snowden. Accessed January 29, 2015. http://www.seas.harvard.edu/news/2015/01/reengineering-privacy-post-snowden.

Phoenix, Krystle. "New Intel Device Could Prevent Hot Car Deaths." *USA Today*. January 16, 2015. Accessed January 17, 2015. http://www.usatoday.com/story/tech/2015/01/16/hot-car-deaths-intel-smart-clip/21849977/.

"Population: Urban, Rural, Suburban." PBS. Accessed April 24, 2015. http://www.pbs.org/fmc/book/1population6.htm.

Poremba, Sue. "Cyber Security Is Growing in Importance for Medical Devices Too." *Forbes*. January 19, 2015. Accessed January 20, 2015. http://www.forbes.com/sites/sungardas/2015/01/19/cyber-security-is-growing-in-importance-for-medical-devices-too/.

Postman, Neil. *Technopoly: The Surrender of Culture to Technology*. New York: Knopf, 1992.

Powell, Barry B. *Writing: Theory and History of the Technology of Civilization*. Chichester, UK: Wiley-Blackwell, 2009.

Pozen, David E. "The Leaky Leviathan: Why the Government Condemns and Condones Unlawful Disclosures Of Information." *Harvard Law Review* 127.512 (2013): 512–635.

Prigg, Mark. "Harvard Professors Warn 'Privacy Is Dead' and Predict Mosquito-Sized Robots that Steal Samples of Your DNA." Daily Mail Online. January 22, 2015. Accessed January 23, 2015. http://www.dailymail.co.uk/sciencetech/article-2921758/Privacy-dead-Harvard-professors-tell-Davos-forum.html.

"Printing Process Explained—Printing History." Dynodan Print Solutions. Accessed May 23, 2015. http://www.dynodan.com/printing-process-explained/printing-history-files/history-lithography.html.

"Privacy Paradox: Teens Don't Perceive Social Media Risk the Way Adults Do." Science 2.0. March 17, 2015. Accessed March 18, 2015. http://www.science20.com/news_articles/privacy_paradox_teens_dont_perceive_social_media_risk_the_way_adults_do-154075.

Putnam, George Haven. *Books and Their Makers during the Middle Ages: A Study of the Conditions of the Production and Distribution of Literature from the Fall of the Roman Empire to the Close of the Seventeenth Century*. New York: Hillary House, 1962.

Quintin, Cooper. "HealthCare.gov Sends Personal Data to Dozens of Tracking Websites." Electronic Frontier Foundation. January 20, 2015. Accessed March 21, 2015. https://www.eff.org/deeplinks/2015/01/healthcare.gov-sends-personal-data.

Ramirez, Edith. "Opening Remarks of FTC Chairwoman Edith Ramirez Privacy and the IoT: Navigating Policy Issues International Consumer Electronics Show." FTC.gov. January 6, 2015. Accessed January 8, 2015. http://www.ftc.gov/system/files/documents/public_statements/617191/150106cesspeech.pdf.

"Records of the British Aviation Industry in the RAF Museum." Royal Air Force Museum. Accessed November 16, 2014. http://www.nationalcoldwarexhibition.org/documents/Guide-to-Aircraft-Industry-Records.pdf.

"The Rialto Police Department's Body-Worn Video Camera Experiment: Operation 'Candid Camera.'" University of Maryland, Department of Criminology and Criminal Justice. April 29, 2013. Accessed March 28, 2015. http://ccjs.umd.edu/sites/ccjs.umd.edu/files/Wearable_Cameras_Capitol_Hill_Final_Presentation_Jerry_Lee_Symposium_2013.pdf.

Rieder, Rem. "Rieder: Ashley Judd's War on Internet Trolls." *USA Today*. March 26, 2015. Accessed March 27, 2015. http://www.usatoday.com/story/money/columnist/rieder/2015/03/25/ashley-judds-campaign-against-virulent-internet-trolls/70432404/.

Riemer, Svend. "Functional Housing in the Middle Ages." Edited by Aaron J. Ihde. *Transactions of the Wisconsin Academy of Sciences, Arts and Letters* XL (1951): 77–91.

Roberts, Donald F., and Ulla G. Foehr. "The Future of Children, Princeton—Brookings: Providing Research and Analysis to Promote Effective Policies and Programs for

Children." The Future of Children. Princeton Brookings. Web. 18 October 2014. http://futureofchildren.org/publications/journals/article/index.xml?journalid=32&articleid=55§ionid=232.

Roberts, Jeff John. "The Internet of Things Is Here, But the Rules to Run It Are Not." Gigaom. 18 October 2014. Web. 19 October 2014. https://gigaom.com/2014/10/18/the-internet-of-things-is-here-but-the-rules-to-run-it-are-not/.

Roberts, Paul. "Vint Cerf: CS Changes Needed to Address IoT Security, Privacy." The Security Ledger. 2 April 2014. Web. 19 October 2014. https://securityledger.com/2014/04/vint-cerf-cs-changes-needed-to-address-iot-security-privacy/.

Robinson, Andrew. *Writing and Script: A Very Short Introduction*. Oxford: Oxford University Press, 2009.

Rogers, Nala. "Israeli Cave Offers Clues about When Humans Mastered Fire." Israeli Cave Offers Clues about When Humans Mastered Fire. December 12, 2014. Accessed December 15, 2014. http://news.sciencemag.org/archaeology/2014/12/israeli-cave-offers-clues-about-when-humans-mastered-fire.

Rogers, Paul. "Streisand's Home Becomes Hit on Web." The Mercury News on Bayarea.com. June 24, 2003. Accessed May 20, 2015. http://www.californiacoastline.org/news/sjmerc5.html

Rogoway, Tyler. "This Is the Army's New Pocket Drone." Foxtrotalpha. November 25, 2014. Accessed November 26, 2014. http://foxtrotalpha.jalopnik.com/this-is-the-armys-new-pocket-drone-1663424760?utm_source=recirculation&utm_medium=recirculation&utm_campaign=wednesdayAM.

Roland, Alex, PhD. "Roland: War & Technology." *American Diplomacy*. American Diplomacy Publishers, Chapel Hill, NC. Web. 22 September 2014. http://www.unc.edu/depts/diplomat/AD_Issues/amdipl_4/roland.html#a.

———. "War and Technology." Foreign Policy Research Institute. http://www.fpri.org/articles/2009/02/war-and-technology. Accessed August 16, 2014.

Rosen, Larry. "iPhone Separation Anxiety." *Psychology Today*. January 18, 2015. Accessed March 6, 2015. https://www.psychologytoday.com/blog/rewired-the-psychology-technology/201501/iphone-separation-anxiety.

Rowan, David. "My Sensors Detect an Internet of Stupid Things." Media News. October 30, 2014. Accessed November 2, 2014. http://www.mediaweek.co.uk/article/1319303/sensors-detect-internet-stupid-things.

Russon, Mary-Ann. "3D Printers Could Be Banned by 2016 for Bioprinting Human Organs." *International Business Times*. 29 January 2014. Web. 21 September 2014. http://www.ibtimes.co.uk/3d-printers-could-be-banned-by-2016-bioprinting-human-organs-1434221.

Ryback, Timothy W. *Hitler's Private Library: The Books that Shaped His Life*. New York: Alfred A. Knopf, 2008.

Sachs, Peter. "Current Drone Law." *Drone Law Journal*. December 14, 2013. Accessed November 9, 2014. http://dronelawjournal.com/.

Saffer, Dan. "The Wonderful Possibilities of Connecting Your Fridge to the Internet | WIRED." Wired.com. Conde Nast Digital, October 29, 2014. Accessed March 20, 2016. http://www.wired.com/2014/10/is-your-refrigerator-running/.

Salzman, Marian. "Google: Privacy? It's So Last Decade." *Forbes*. February 27, 2015. Accessed February 28, 2015. http://www.forbes.com/sites/mariansalzman/2015/02/27/google-privacy-its-so-last-decade/.

Sante, Luc. *Low Life: Lures and Snares of Old New York*. New York: Farrar, Straus and Giroux, 1991.

"SatNav Danger Revealed: Navigation Device Blamed for Causing 300,000 Crashes." *Mirror*. 21 July 2008. Web. 14 September 2014. http://www.mirror.co.uk/news/uk-news/satnav-danger-revealed-navigation-device-319309.

Scarre, Christopher, ed. *The Human Past: World Prehistory & the Development of Human Societies*. London: Thames & Hudson, 2005.

Scheer, Robert. "The Internet Killed Privacy: Our Liberation, and Our Capture, Are within the Same Tool." Saloncom RSS. March 7, 2015. Accessed March 8, 2015. http://www.salon.com/2015/03/07/the_internet_killed_privacy_our_liberation_and_our_capture_are_within_the_same_tool/.

Schmandt-Besserat, Denise. *Before Writing*, Vol. 1. 1st ed. Austin: University of Texas Press, 1992.

———. *Before Writing*, Vol. 2. 1st edAustin: University of Texas Press, 1992.

———. *How Writing Came About*. Austin: University of Texas Press, 1996.

———. "The Earliest Precursor of Writing." *Scientific American*, June 1, 1977, 50–58.

———. "Two Precursors of Writing: Plain and Complex Tokens." Escola Finaly. In *The Origins of Writing*, Wayne M. Senner, Ed. December 8, 2008. Accessed December 7, 2014. http://en.finaly.org/index.php/Two_precursors_of_writing:_plain_and_complex_tokens.

Schmidt, Eric, and Jared Cohen. *The New Digital Age: Reshaping the Future of People, Nations and Business*. New York: Vintage / Random House, 2013.

Schnaars, Steven P., and Sergio Carvalho. "Predicting the Market Evolution of Computers: Was the Revolution Really Unforeseen?" *Technology in Society* 26 (2004): 1–16.

Schneier, Bruce. "The Big Idea." Whatever: All Cake and Hand Grenades. March 4, 2015. Accessed March 16, 2015. http://whatever.scalzi.com/2015/03/04/the-big-idea-bruce-schneier-2/.

———. "Surveillance as a Business Model." Schneier on Security. 1 November 2013. Web. 6 October 2014. https://www.schneier.com/blog/archives/2013/11/surveillance_as_1.html.

———. "Schneier on Security." Blog. Schneier on Security, 9 April 2014. Web. 11 October 2014. https://www.schneier.com/news/archives/2014/04/surveillance_is_the.html.

———. "The Equation Group's Sophisticated Hacking and Exploitation Tools." Blog. Schneier on Security. February 17, 2015. Accessed February 18, 2015. https://www.schneier.com/blog/archives/2015/02/the_equation_gr.html.

———. "Your TV May Be Watching You." CNN. February 12, 2015. Accessed February 16, 2015. http://www.cnn.com/2015/02/11/opinion/schneier-samsung-tv-listening/index.html.

Scott, Alwyn. "U.S. OK's More Commercial Drone Use as Congress Probes Risks." Reuters. December 10, 2014. Accessed December 15, 2014. http://www.reuters.com/article/2014/12/10/us-usa-drones-faa-idUSKBN0JO1J920141210.

"Secret Life of the Cat: The Science of Tracking Our Pets." BBC News. BBC, 12 June 2013. Web. 24 August 2014. http://www.bbc.com/news/science-environment-22821639.

"Security Advisories for Firefox." Mozilla. March 31, 2015. Accessed April 1, 2015. https://www.mozilla.org/en-US/security/known-vulnerabilities/firefox/.

Shaban, Hamza. "Will the Internet of Things Finally Kill Privacy?" The Verge. February 11, 2015. Accessed February 12, 2015. http://www.theverge.com/2015/2/11/8016585/will-the-internet-of-things-finally-kill-privacy.

Shandrow, Kim LaChance. "Jibo, the Personal Robot Startup, Lands $25 Million in Funding." *Entrepreneur*. January 21, 2015. Accessed January 23, 2015. http://www.entrepreneur.com/article/242047.

Shaver, Katherine. "Experts: Using Smartphones While Walking Is a Hazard." *Virginian-Pilot*. September 21, 2014. Accessed March 3, 2015. http://hamptonroads.com/2014/09/experts-using-smartphones-while-walking-hazard.

Shaw, Ian G. R. "History of U.S. Drones." The Rise of the Predator Empire: Tracing the History of U.S. Drones. 2013. Accessed November 29, 2014. https://understandingempire.wordpress.com/2-0-a-brief-history-of-u-s-drones/.

Shelton, LeeAnn. "Las Vegas Man Arrested for Naperville 'Swatting' Incident." *Chicago Sun-Times*. February 6, 2015. Accessed March 20, 2015. http://chicago.suntimes.com/crime/7/71/350553/las-vegas-man-arrested-naperville-swatting-incident.

Sides, Hampton. *In the Kingdom of Ice: The Grand and Terrible Polar Voyage of the USS* Jeannette. New York: Doubleday, 2014.

Simonite, Tom. "Air Traffic Control for Drones." *MIT Technology Review*. October 17, 2014. Accessed December 12, 2014. http://www.technologyreview.com/news/531811/air-traffic-control-for-drones/.

"Sir Timothy Berners-Lee Interview." Academy of Achievement. January 22, 2007. Accessed April 7, 2015. http://www.achievement.org/autodoc/page/ber1int-1.

"Skully Augmented Reality Crash Helmet Heading to Intermot." Tech News | The Star Online. The Star Online, 30 September 2014. Web. 30 September 2014. http://www.thestar.com.my/Tech/Tech-News/2014/09/30/Skully-Augmented-reality-crash-helmet-heading-to-Intermot/.

Smil, Vaclav. *Transforming the Twentieth Century: Technical Innovations and Their Consequences*. Oxford: Oxford UP, 2006.

Sobel, Dava. *Longitude: The True Story of a Lone Genius Who Solved the Greatest Scientific Problem of His Time*. New York: Walker, 1995.

Sonn, William J. *Paradigms Lost: The Life and Deaths of the Printed Word*. Lanham, MD: Scarecrow Press, 2006.

"South Tyrol Museum of Archaeology." Museum Home. Web. 8 August 2014. http://www.iceman.it/en.

Spagat, Elliot, and Brian Skoloff. "AP Exclusive: Drones Patrol Half of Mexico Border." ABC News. November 12, 2014. Accessed November 17, 2014. http://abcnews.go.com/US/wireStory/ap-exclusive-drones-patrol-half-mexico-border-26877520.

"Spitting Image." *The Economist: Technology Quarterly*, September 19, 2002.

Srinivasan, Sriram, and Sangeetha Kandavel. "Facebook Is a Surveillance Engine, Not Friend: Richard Stallman, Free Software Foundation." *Times of India—Economic*

Times. February 7, 2012. Accessed March 23, 2015. http://articles.economictimes.indiatimes.com/2012-02-07/news/31034052_1_facebook-users-mark-zuckerberg-richard-stallman.

Stafford, Dave. "$1.4M Judgment against Walgreen Sets Health Privacy Precedent." *Indianapolis Business News*. November 15, 2014. Accessed November 16, 2014. http://www.ibj.com/articles/50510-14m-judgment-against-walgreen-sets-health-privacy-precedent.

Stanley, Jay. "Police Body-Mounted Cameras: With Right Policies in Place, a Win for All." American Civil Liberties Union. March 1, 2013. Accessed March 25, 2015. https://www.aclu.org/technology-and-liberty/police-body-mounted-cameras-right-policies-place-win-all.

"State of Metropolitan America: On the Front Lines of Demographic Transformation." The Brookings Institution. January 1, 2010. Accessed April 24, 2015. http://www.brookings.edu/~/media/research/files/reports/2010/5/09-metro-america/metro_america_report.pdf.

Sterling, Bruce. *The Epic Struggle of the Internet of Things*. Moscow & London: Strelka, 2014.

Sterns, Olivia. "Muslim Inventions that Shaped the Modern World." CNN. January 29, 2010. Accessed May 10, 2015. http://www.cnn.com/2010/WORLD/meast/01/29/muslim.inventions/.

Stockley, Mark. "How Nine out of Ten Healthcare Pages Leak Private Data." Naked Security. February 26, 2015. Accessed February 27, 2015. https://nakedsecurity.sophos.com/2015/02/26/how-nine-out-of-ten-healthcare-pages-leak-private-data/.

———. "The Dark Web: Anarchy, Law, Freedom and Anonymity." Naked Security. February 20, 2015. Accessed February 20, 2015. https://nakedsecurity.sophos.com/2015/02/20/the-dark-web-anarchy-law-freedom-and-anonymity/.

Stoll, Clifford. "Why the Web Won't Be Nirvana." *Newsweek*. 26 February 1995.

Sullivan, Mark. "NFC Might Turn Apple's iWatch into Your Personal Passkey to the World." VentureBeat. 4 September 2014. Web. 5 September 2014. http://venturebeat.com/2014/09/04/nfc-might-turn-apples-iwatch-into-your-personal-passkey-to-the-world/.

Taylor, Bob. "Charles E. Taylor: The Man Aviation History Almost Forgot." Federal Aviation Administration. Accessed November 19, 2014. https://www.faa.gov/about/office_org/field_offices/fsdo/phl/local_more/media/CT Hist.pdf.

Taylor, Heather. "Cost of Constructing a Home." Housing Economics.com. National Association of Home Builders, 2 January 2014. Web. 14 September 2014. http://www.nahb.org/generic.aspx?sectionID=734&genericContentID=221388&channelID=311.

Temple, James. "Facebook IPO Underscores Shutting out the Masses." SFGate. May 22, 2012. Accessed November 16, 2014. http://www.sfgate.com/business/article/Facebook-IPO-underscores-shutting-out-the-masses-3575283.php.

Tiku, Natasha, and Casey Newton. "Twitter CEO: 'We Suck at Dealing with Abuse'" The Verge. February 4, 2015. Accessed February 5, 2015. http://www.theverge.com/2015/2/4/7982099/twitter-ceo-sent-memo-taking-personal-responsibility-for-the.

Timberg, Craig. "Is Uber's Rider Database a Sitting Duck for Hackers?" *Washington Post*. December 1, 2014. Accessed January 24, 2015. http://www.washingtonpost.com/blogs/the-switch/wp/2014/12/01/is-ubers-rider-database-a-sitting-duck-for-hackers/.

"Timeline of Computer History." Computer History Museum. Accessed April 5, 2015. http://www.computerhistory.org/timeline/?category=cmptr.

Titcomb, James. "Google Buys Drone Manufacturer Titan Aerospace." *The Telegraph*. April 14, 2014. Accessed November 16, 2014. http://www.telegraph.co.uk/technology/google/10766490/Google-buys-drone-manufacturer-Titan-Aerospace.html.

Todd, Deborah M. "Report: Privacy Policies Still Stump Americans." *Pittsburgh Post-Gazette*. November 26, 2014. Accessed November 26, 2014. http://www.post-gazette.com/business/tech-news/2014/11/26/Report-Privacy-policies-still-stump-Americans/stories/201411260123.

"Top Ten Uses for a Mobile Phone? Calls Come SIXTH! 40% of Smartphone Users Say They Could Manage without Call Function on Their Device." Mail Online. October 30, 2014. Accessed March 8, 2015. http://www.dailymail.co.uk/news/article-2815114/.

"Tracking & Hacking: Security & Privacy Gaps Put American Drivers at Risk." Ed Markey: US Senator for Massachusetts. Accessed February 9, 2015. http://www.markey.senate.gov/imo/media/doc/2015-02-06_MarkeyReport-Tracking_Hacking_CarSecurity.pdf.

Trowbridge, Alexander. "Evolution of the Phone: From the First Call to the Next Frontier." CBSNews. December 16, 2014. Accessed March 5, 2015. http://www.cbsnews.com/news/evolution-of-the-phone-from-the-first-call-to-the-next-frontier/.

"Truman Refuses to Rule out Atomic Weapons." History.com. Accessed March 31, 2015. http://www.history.com/this-day-in-history/truman-refuses-to-rule-out-atomic-weapons.

Turse, Nick. *The Changing Face of Empire Special Ops, Drones, Spies, Proxy Fighters, Secret Bases, and Cyberwarfare*. New York: Haymarket, 2012.

"Twitter Usage Statistics." Internet Live Stats. Internetlivestats.com. Web. 7 October 2014. http://www.internetlivestats.com/twitter-statistics/.

Twomey, Terrence. "How Domesticating Fire Facilitated the Evolution of Human Cooperation." *Biology & Philosophy*, February 2013, 89–99.

———. "Keeping Fire: The Cognitive Implications of Controlled Fire Use by Middle Pleistocene Humans." PhD dissertation, School of Social and Political Sciences, Faculty of Arts, The University of Melbourne, 2011.

———. "The Cognitive Implications of Controlled Fire Use by Early Humans." *Cambridge Archaeological Journal* 23, 01 (2013): 113–28.

"Typhoon Haiyan (Yolanda)." US Agency for International Development. United States Government. Web. 12 October 2014. http://www.usaid.gov/haiyan.

"Typhoon Haiyan Toll Rises over 5,000." BBC News. British Broadcasting Corporation, 22 November 2013. Web. 12 October 2014. http://www.bbc.com/news/world-asia-25051606.

UC Berkeley. "Berkeley Robotics & Human Engineering Laboratory." Berkeley Robotics Human Engineering Laboratory RSS. http://bleex.me.berkeley.edu/research/exoskeleton/hulc/ (accessed September 25, 2014).

"US Repeatedly Threatened to Use Nukes on N. Korea: Declassified Documents." Raw Story / Associated Press. October 9, 2010. Accessed April 10, 2015. http://www.rawstory.com/rs/2010/10/repeatedly-threatened-nukes-korea-declassified-documents/.

Vaas, Lisa. "Breakthrough in Facial Recognition: The 'Deep Dense Face Detector.'" Naked Security. February 19, 2015. Accessed February 19, 2015. https://nakedsecurity

.sophos.com/2015/02/19/breakthrough-in-facial-recognition-the-deep-dense-face-detector/.

———. "Entire Oakland Police License Plate Reader Data Set Handed to Journalist." Naked Security. March 26, 2015. Accessed March 26, 2015. https://nakedsecurity.sophos.com/2015/03/26/entire-oakland-police-license-plate-reader-data-set-handed-to-journalist/.

———. "Gamer Swatted While Live-Streaming on Twitch TV." Naked Security. February 10, 2015. Accessed February 10, 2015. https://nakedsecurity.sophos.com/2015/02/10/gamer-swatted-while-live-streaming-on-twitch-tv/.

———. "Internet of Things Is a Threat to Privacy, Says FTC." Naked Security. January 8, 2015. Accessed January 8, 2015. https://nakedsecurity.sophos.com/2015/01/08/internet-of-things-is-a-threat-to-privacy-says-ftc/.

———. "Twitter Troll Fired, Another Suspended after Curt Schilling Names and Shames Them." Naked Security. March 5, 2015. Accessed March 5, 2015. https://nakedsecurity.sophos.com/2015/03/05/twitter-troll-fired-another-suspended-after-curt-schilling-names-and-shames-them/.

———. "Uber Goes Big Data, Shares Customers' Data with a Hotel Chain." Naked Security. March 26, 2015. Accessed March 26, 2015. https://nakedsecurity.sophos.com/2015/03/26/uber-goes-big-data-shares-customers-data-with-a-hotel-chain/.

———. "YouTube Channel Swamps Police with Requests for Disclosure of Body-Cam Video." Naked Security. November 13, 2014. Accessed November 13, 2014. https://nakedsecurity.sophos.com/2014/11/13/youtube-channel-swamps-police-with-requests-for-disclosure-of-body-cam-video/.

Van Creveld, Martin. *Technology and War: From 2000 B.C. to the Present*. New York: Touchstone, 1991.

Vincent, James. "MIT's 'Immersion' Project Reveals the Power of Metadata." *The Independent*. Independent Digital News and Media, 8 July 2013. Web. 12 October 2014. http://www.independent.co.uk/life-style/gadgets-and-tech/mits-immersion-project-reveals-the-power-of-metadata-8695195.html.

Violino, Bob. "The 'Internet of Things' Will Mean Really, Really Big Data." *InfoWorld*. 29 July 2013. Web. 25 October 2014. http://www.infoworld.com/article/2611319/computer-hardware/the—internet-of-things—will-mean-really—really-big-data.html.

Vizard, Mike. "Mayo Clinic Uses 3D Printer to Create Customized Artificial Hip." 3D Printer World. 3D Printer World, 15 April 2013. Web. 21 September 2014. http://www.3dprinterworld.com/article/mayo-clinic-uses-3d-printer-create-customized-artificial-hip.

Walker, Steven, Vasa Lukich, and Michael Chazan. "Kathu Townlands: A High Density Earlier Stone Age Locality in the Interior of South Africa." San Francisco: PLoS ONE 9(7): E103436 (PLoS.org), 2014.

Walker, Williston. *A History of the Christian Church*. New York: Scribner, 1959.

Webb, Geoff. "Say Goodbye to Privacy." Wired.com. February 5, 2015. Accessed February 6, 2015. http://www.wired.com/2015/02/say-goodbye-to-privacy/.

Weinberg, Michael. "What's the Deal with Copyright and 3D Printing?" /www.publicknowledge.org. Public Knowledge. Web. 8 August 2014. http://tutorial8.com/w/whats-the-deal-with-copyright-and-3d-printing-w25752-pdf.pdf.

Weinstein, Mark. "5 Newsworthy Privacy Stories Today." *Huffington Post*. March 10, 2015. Accessed March 13, 2015. http://www.huffingtonpost.com/mark-weinstein/5-news worthy-privacy-stor_b_6834940.html.

Weisman, Steve. "Are You Safe in the Internet of Things?" *USA Today*. April 4, 2015. Accessed April 7, 2015. http://www.usatoday.com/story/money/columnist/2015/ 04/04/weisman-internet-of-things-cyber-security/70742000/.

Weissenbacher, Manfred. *Sources of Power: How Energy Forges Human History*. Santa Barbara, CA: Praeger, 2009.

Wertheim, Margaret. "Robots that Build (But Still Won't Do Windows)." *New York Times*. Accessed September 28, 2014. http://www.nytimes.com/2004/03/11/garden/robots -that-build-but-still-won-t-do-windows.html.

"What Happens When the Internet Goes Out? This Arizona Town Found Out." *Los Angeles Times*. Associated Press. February 26, 2015. Accessed March 17, 2015. http://www .latimes.com/business/la-fi-arizona-internet-outage-20150227-story.html.

Whittle, Richard. "The Man Who Invented the Predator." *Air & Space* Magazine. April 2013. Accessed November 29, 2014. http://www.airspacemag.com/flight-today/the -man-who-invented-the-predator-3970502/?no-ist.

"Who Has Your Back? Government Data Requests 2014." Electronic Frontier Foundation. Accessed November 4, 2014. https://www.eff.org/who-has-your-back-government -data-requests-2014.

"Why Is Sitting by a Fire so Relaxing? Evolution May Hold the Key." Seriously Science. Accessed November 13, 2014. http://blogs.discovermagazine.com/seriouslyscience/ 2014/11/13/evolutionary-explanation-sitting-fire-relaxing/.

Wiessner, Polly. "Embers of Society: Firelight Talk among the Ju/'hoansi Bushmen." *Proceedings of the National Academy of Sciences* 111, no. 39 (2014): 14027–4035.

Willingham, Daniel T. "Smartphones Don't Make Us Dumb." *New York Times*. January 20, 2015. Accessed January 21, 2015. http://www.nytimes.com/2015/01/21/opinion/ smartphones-dont-make-us-dumb.html.

Wilson, Reid. "Wyoming Will Debate Digital Privacy Amendment." *Washington Post*. November 17, 2014. Accessed November 18, 2014. http://www.washingtonpost.com/ blogs/govbeat/wp/2014/11/17/wyoming-will-debate-digital-privacy-amendment/.

Wolmar, Christian. *The Iron Road: An Illustrated History of the Railroad*. New York: DK, 2014.

Woods, Ben. "How to Use Personal Drones Legally." TNW Network All Stories RSS. July 4, 2014. Accessed November 9, 2014. http://thenextweb.com/gadgets/2014/07/04/ use-personal-drones-legally-beginners-guide/.

"World Bank Group." "Mobile Cellular Subscriptions," World Bank Group. Accessed March 3, 2015. http://www.worldbank.org/.

"World Fire Statistics." *Geneva Association: Bulletin* 29 (2014): 5–7.

Worstall, Tim. "The Story of Henry Ford's $5 a Day Wages: It's Not What You Think." *Forbes*. March 12, 2012. Accessed April 21, 2015. http://www.forbes.com/sites/tim worstall/2012/03/04/the-story-of-henry-fords-5-a-day-wages-its-not-what-you -think/.

Wrangham, Richard W., and Dale Peterson. *Demonic Males: Apes and the Origins of Human Violence*. Boston: Houghton Mifflin, 1996.

"The Wright Story / The Airplane Business." The Wright Story. Accessed November 16, 2014. http://www.wright-brothers.org/History_Wing/Wright_Story/Airplane_Business/Airplane_Business_Intro.htm.

Zaharov-Reutt, Alex. "Gartner: Smartphones Sales Top 1.2 Billion in 2014, Apple No.1." ITwire. March 4, 2015. Accessed March 4, 2015. http://www.itwire.com/it-industry-news/telecoms-and-nbn/67171-gartner-smartphones-sales-top-12-billion-in-2014-apple-no1.

Zetter, Kim. "A Cyberattack Has Caused Confirmed Physical Damage for the Second Time Ever." Wired.com. January 1, 2015. Accessed January 17, 2015. http://www.wired.com/2015/01/german-steel-mill-hack-destruction/.

Zorabedian, John. "Visa Asks to Track Your Smartphone to Help Sniff out Credit Card Fraud." Naked Security. February 18, 2015. Accessed February 19, 2015. https://nakedsecurity.sophos.com/2015/02/18/visa-asks-to-track-your-smartphone-to-help-sniff-out-credit-card-fraud/.

Zorich, Zach. "From the Trenches—New Evidence for Mankind's Earliest Migrations—*Archaeology* Magazine Archive." March 1, 2011. Accessed January 13, 2015. http://archive.archaeology.org/1105/trenches/homo_sapien_migration_africa_arabia.html.

Zuckerman, Sam. "As Internet Surges, Data Centers Gear Up." *USA Today*. Gannett, 25 October 2014. Web. 27 October 2014. http://www.usatoday.com/story/tech/2014/10/25/internet-surges-data-centers-gear-up/17381833/.

Interviews and Correspondence

Agbodjan, Edoe. E-mail and face-to-face interviews by author. Lincoln, Nebraska, July 15, 2014. Ottowa, Ontario, Canada, August 2, 2014.

Angelini, Chris. E-mail interview by author. Bakersfield, California, November 9, 2014.

Bock, Michael. Daimler AG. E-mail interview by author. Stuttgart, Germany, May 22, 2015.

Bowyer, Dr. Adrian. E-mail interview by author. Bath, UK, June 27, 2014.

boyd, danah. E-mail interview by author. New York, New York, March 12, 2015.

Brown Kramer, Josh. E-mail interview by author. Lincoln, Nebraska, July 12, 2014.

Browne, Andrew. E-mail interview by author. San Francisco, California, August 15, 2014.

Buhler, Stephen, PhD. University of Nebraska. E-mail interview by author. Lincoln, Nebraska, May 28, 2015.

Campos-Castillo, Celeste. E-mail interview by author. Milwaukee, Wisconsin, December 12, 2014 and March 11, 2015.

Cardwell, Clay. E-mail interview by author. Omaha, Nebraska, August 5, 2014.

Christensen, Brett. E-mail interview by author. Bundaberg, Queensland, Australia, October 1, 2014.

ChuanJun, Yang. E-mail interview by author. Suzhou, Jiangsu Province, China, June 22, 2014.

Davidson, Iver. E-mail interview by author. Las Cruces, New Mexico, September 24, 2014.

Denton, Chad. E-mail interview by author. Brooklyn, New York, September 9, 2014.

Dyer, Shane. Interview by author. Phone interview. Lincoln, Nebraska, and Redwood City, California, July 17, 2014.

Eisen, Joe. Interview by author. Phone interview. London, England, August 18, 2014.

Enderle, Rob. Skype video interview by author. San Jose, California, March 27, 2015.

Froom, Robert Keith. E-mail interview by author. Garner, North Carolina, October 8, 2014.
Golden, Diane. E-mail interview by author. Kansas City, Missouri, July 12, 2014.
Grady, David. Interview by author. Personal interview. Lathrop, Missouri, July 20, 2014.
Hadnagy, Chris. Skype video interview by author. Philadelphia, Pennsylvania, September 7, 2014.
Hess, Sean. Skype video interview by author. Boise, Idaho, September 7, 2014.
Kenealy, Tim. E-mail interview by author. Minneapolis, Minnesota, April 13, 2014.
Kirkbride, Ryan, PhD. Skype video interview by author. Austin, Texas, August 10, 2014 and September 7, 2014.
Lawson, Agnele. E-mail interview by author. Paris, France, August 15, 2014.
Lifton, Josh. E-mail interview by author. Portland, Oregon, April 4, 2015.
Mawston, Neil. E-mail interview by author. Milton Keynes, UK, April 17, 2014.
Nemat-Nasser, Syrus. Interview by author. E-mail interview. San Diego, California, July 11, 2014.
O'Gorman, Jim. E-mail interview by author. Charlotte, North Carolina, September 24, 2014.
Oppenheim, Casey. E-mail interview by author. Palo Alto, California, September 1, 2014.
Peed, Shane. Interview by author. Personal interview. Lincoln, Nebraska, May 13, 2014.
Perry, Michael. E-mail interview by author. Los Angeles, California, July 28, 2014 and September 1, 2014.
Quinn, Zoe, and Alex Lifschitz. Skype video interview by author, January 30, 2015. (Location redacted.)
Ratner, Nan Bernstein. Professor, Department of Hearing and Speech Sciences, University of Maryland. E-mail interview by author. College Park, Maryland, December 16, 2014.
Sagona, Antonio. E-mail interview by author. Melbourne, Victoria, Australia, December 9, 2014 and February 8, 2015.
Sanwal, Anand. E-mail interview by author. New York, New York, October 19, 2014.
Schneider, Daniel. University of Michigan Health System. E-mail interview by author. Ann Arbor, Michigan, January 26, 2015.
Schneier, Bruce. E-mail interview by author. Cambridge, Massachusetts, October 12, 2014.
Swannie, Karl. Telephone interview by author. Victoria, British Columbia, Canada, October 3, 2014.
Tone, Nathan. Interview by author. Phone interview. Kentfield, California, August 12, 2014.
Trevino, Jorge. E-mail interview by author. Monterrey, Mexico, August 26, 2014.
Twomey, Terrence. E-mail interview by author. Melbourne, Victoria, Australia, December 18, 2014 and May 15, 2015.
Warde, Jeff. E-mail interview by author. Longmont, Colorado, August 18, 2014.
Welch, David. Telephone interview by author. Saffron Walden, UK, October 22, 2014.
Wikimedia Foundation Spokesperson. E-mail interview by author. San Francisco, California, April 30, 2015.
Wilcox, Dulcie. E-mail interview by author. Longmont, Colorado, April 18, 2014.
York, Jillian. E-mail interview by author. Berlin, Germany. March 6, 2015.

Index

X

Z